MATHEMATICS CONTESTS

A Guide for Involving Students and Schools

Frederick O. Flener
Northeastern Illinois University

NATIONAL COUNCIL OF TEACHERS OF MATHEMATICS

Copyright © 1990 by
THE NATIONAL COUNCIL OF TEACHERS OF MATHEMATICS, INC.
1906 Association Drive, Reston, Virginia 22091
All rights reserved.

Library of Congress Cataloging-in-Publication Data:

Flener, Frederick O.
    Mathematics contests : a guide for involving students and schools
/ Frederick O. Flener.
        p. cm.
    ISBN 0-87353-282-1
    1. Mathematics—Competitions—Handbooks, manuals, etc.
    2. Mathematics—Examinations, questions, etc.   I. Title.
    QA99.F64 1989
    510'.79—dc20   89–28855
                                                                CIP

Printed in the United States of America

Contents

Preface

As a coach, question writer, and general consultant to several mathematics leagues, I have been actively involved in mathematics contests for more than twenty years. The contest formats represented in this booklet are those with which I have experience, but certainly many others exist.

All the questions given here are those that I have written for various mathematics contests. I wish I could say with certainty that all are original, but many of the questions have been derived from other sources, such as workshops, conferences, cocktail parties, and even former students. Therefore, a reader may see variations of his or her own questions in this booklet. All the questions are expressed in my own words. For example, a problem may originally have stated, "What is the shortest distance for an ant to walk along the surface of a cube from one vertex to a vertex at the opposite end of the diagonal?" I tried to rewrite this question slightly by saying, "Argyle the ant, living at one vertex, wants to visit his girlfriend Beatrice. . . ."

It is impossible to cite all the sources for the (literally) thousands of questions I have written; I can only say, "Thank you." I hope readers find this booklet a useful source for organizing and conducting mathematics contests for the benefit and enjoyment of many high school students.

Introduction

Why Have a Mathematics Contest?

Competition has social implications for all students. For students who have difficulty learning mathematics, every test, every quiz, and every homework assignment can be a source of frustration that leads to a poor self-image. A math contest, however, with the team spirit fostered by studying for the competition and an atmosphere free of the grading system, may be the challenge and reinforcement that low-achieving students need.

Students who are successful in the classroom and are rarely challenged intellectually can also benefit from a contest. Participating can motivate them to higher levels of achievement than they would reach in their regular classes. These students are disappointed after a defeat, but they investigate their mistakes and use the experience to improve their competence.

Consider the following: In the State of Illinois Mathematics Contest, contestants in various categories answer an average of 40–60 percent of the questions *incorrectly,* but the number of participants in the contest is increasing each year. Even students who answer correctly only five questions out of twenty compete again the following year, possibly with an even stronger motivation to do well. In fact, the difficulty of the questions appears to be increasing each year because the contestants are coming to the contests better prepared. The contests give students the opportunity to learn from their mistakes without being penalized by the lower grades that would result from mistakes on classroom tests.

Another intrinsic benefit is the contests' potential to improve a teacher's attitude toward teaching. Teaching tends to be a rewarding profession, but after five classes a day for 180 days a year, even the most rewarding job can become routine. Mathematics contests are far from routine. A teacher who helps prepare the students for a contest has an opportunity to explore problems and topics that are beyond the scope of the normal classroom.

To attain maximum teacher benefit from competition, keep in mind that participation by several faculty members with small responsibilities is far superior to having one or two teachers take on complete responsibility. Although some teachers are initially reluctant to participate, it is worthwhile encouraging them. A second note of caution: Teachers should not come to practice sessions knowing how to solve all the problems. Nothing could reduce the stimulation of the contest for the teacher as quickly as thinking of the practice session as an additional class preparation. Furthermore, it is

1

the students who need to solve the problems, not the teacher, although the teacher and the students can work together. These techniques will be elaborated on in the section about coaching strategies. If handled correctly, the contests can be stimulating to both students and teachers.

Possibly the most positive outcome of participating in mathematics contests is the potential for changing the image of mathematics from something austere and intellectual to something that is sociable and enjoyable. Students can have T-shirts, luncheons, parties, and school recognition. In many schools a mathematics club becomes more popular simply because the preparation for contests gives the students a focal point for communication. The students can help one another (and should be encouraged to do so). It would not be uncommon for twelfth-grade students to befriend ninth graders. In one school the "mathletes" dinner party gained a social status topped only by the traditional events of homecoming and the prom.

Before You Start

The benefits described in the preceding section will not spring up spontaneously like dandelions. But preparations are not very difficult, and once started, a mathematics contest program will often move under its own momentum. In this section several suggestions will be made to help initiate a program.

At least one person has to become the head coach and take the initiative for starting a contest program. It should not be started, however, without support from other faculty, the administration, and even the board of education. It is probably not necessary to have a full commitment from the students, but only to be sure the students are capable of competing with some degree of respectability. The students can usually be nudged into participation, and once started, it will not be difficult to maintain their participation.

The support from other faculty is necessary because they have to help, at least on a limited basis, by coaching students and monitoring contests hosted by the school. The head coach should approach his or her colleagues with a general plan that describes their limited participation. For example, a coach might say, "I need someone to work with five or six students for the second contest. The topic will be matrices, and I'll give the kids all the problems to work on. All you have to do is meet with them three or four times after school for about an hour." This might be the annual commitment of one faculty member. Others might be willing to take on three or four such tasks. Procedures for using the faculty will be discussed in the section on coaching.

The administration can undermine the program by a lack of support. For example, most contest programs require fees for participation, transportation for students, the use of the facilities to host a contest, and possibly even the release of students from school to attend contests. Unless the administration understands the extent of the commitment the school must make and agrees to it, full participation will be difficult.

The real source of recognition should be the board of education. If the administration wants to have the program, they will carry the case to the board. More likely, the administration will support your effort but may be unable to develop a rationale for the program. When the case for participation in math contests is presented to the board, do not be timid. A mathematics team is as beneficial to the students as any other competitive team. Ask the board, "Does a state champion in mathematics deserve less acclaim than, say, the state champion in tennis?" At a public meeting, how can any board member say no? The team and the coaches need to have public recognition as a foundation to start the program.

Should the coach of the mathematics team be paid? Probably so, particularly if other coaches are paid. It is unlikely that the addendum a coach receives will be a stimulus to attract a teacher to the job, but at least there should be recognition. Nothing dampens enthusiasm more than a belief that one is being taken advantage of. Therefore, work done with the mathematics team should be compensated similarly to the coaching for other extracurricular activities. One should not take on the coaching job solely for the extra stipend. The dollar rewards will likely be minor on a per-hour basis. Rather, the stipend should be considered as a recognition for one's effort.

Getting Started

Once the faculty, administration, and the board of education are convinced that students should compete in mathematics, it is time to get ready. But how do you start? With whom do you compete?

Contests can be organized at many levels—intramural, regional, statewide, national, or even international. There are many existing contest organizations or leagues. The National Council of Teachers of Mathematics has published *Mathematics Contests: A Handbook for Mathematics Educators* (1982), which lists 123 competitions throughout the United States. This is only a small fraction of the contests that are conducted each year.

The level of participation for these existing organizations, as well as the contest formats, varies considerably. Many of the contests are restricted to local schools, but some, such as the Atlantic-Pacific High School Mathematics League, are open to any high school in the United States. Some contests have very simple formats so that any number of students can compete, and others have very complex formats for which each team must have thirty to forty members to compete in full.

A school that is starting to compete has three options: to join an existing organization, to convince other schools to begin a competitive program, or simply to have the students compete among themselves. Possibly the easiest route is to join an existing league; the most difficult, but maybe the most rewarding, is to get other schools to begin a program for meeting in head-to-head competition.

A problem in joining an existing league at the local or national level is that the other schools may have much more experience; new participants can quickly become discouraged, and only the highly talented will receive any recognition. Therefore, starting your own competition program, either between schools with similar student populations or within your school, may be the best way to begin. Several decisions must be made before organizing your own contest:

- Extent of student participation
- Contest format
- Distribution of organizational responsibilities

The first consideration in organizing the contest is to determine who will participate. Do you want to limit the contest to students in the upper classes? Do you want to have a limited number of students participating, or is the objective to have a large number of students competing? Will you expect the students to prepare for specific topics, or will there be tests of general mathematical aptitude? Will the students work independently, or will there be some provision for group work? This discussion of the extent of student participation is crucial, because there are many formats that can be used in the contests, and the selection should be based on the type and number of students participating.

If the contests are limited to only a few of the best students at each school, then the questions should be challenging and the solutions may be open-ended, as in a proof. If the contest will be open to forty to fifty students from each school, then the questions need to be graded quickly and there may be a variety of formats, such as in a field day. If there is some emphasis on group work, then relays, two-person teams, eight-person teams, and so on, should be used. These various formats, including sample questions, will be described later.

Once the format is decided on, it is important to have specific responsibilities assigned to the faculty in the competing schools. There are some important considerations prior to having a contest.

Facilities. Rooms must be available for competitions, for the schools' teams to meet informally, for grading exams, and for the scores to be reported, and there should be a place for the coaches to have refreshments.

Materials. Someone must prepare the questions for the contests and reproduce all the copies that are needed. Often the questions are prepared and reproduced by an outside source. Obviously, the cost must be shared by participating schools, but this may be better than having faculty in the competing schools prepare the questions. Someone must also provide the awards, which may be as routine as ribbons or certificates or more substantive, such as calculators or scholarships.

Personnel. Necessary personnel will include proctors, graders, a protest committee, someone responsible for awards, and someone to keep records and statistics. These responsibilities should be shared by all the participating schools.

Suggestions for Coaching

As stated earlier, an effective mathematics contest program requires the support and involvement of many school administrators and faculty, but there are one or two persons who are primarily responsible for coaching the team. The responsibilities of the coaches will vary considerably, depending on the format of the contest program.

One important role of a coach is often to participate in some type of central governing group. Unlike baseball or basketball, the rules for mathematics contests are flexible and open to continuous modifications. Therefore, participating schools need to meet occasionally to discuss the contests. In many ways the work of this central group resembles a political system. There are the "conservatives," who wish to limit the participation and to maintain a high-quality program, and there are the "liberals," who prefer innovative formats and esoteric topics with wide open participation. Through some type of democratic process, the contest program will eventually become a compromise of many viewpoints.

A coach should not take his or her role in this group lightly. Each school's viewpoint must be represented in determining the contest format. If the format selected puts a school at a distinct disadvantage, then it may be difficult to maintain the students' enthusiasm for the contests. For example, if the format requires thirty-five to forty students for a full team, then a school that can expect to recruit only twenty would be unable to achieve any school awards. Although individuals from that school can be successful, it is team success that keeps the program going. If the school is successful, then other students will want to join the team, the administration will believe the program is accomplishing worthwhile objectives, and the school board will love the publicity.

The main task of the coach will be to manage the program within the school. To prepare for a contest may be as simple as selecting a few students to take a test at a single sitting, but many contests are far more complicated than that. In most instances the contests will involve students from all grade levels, and there will be a variety of categories. Having other faculty members to help will be a tremendous asset in preparing students for the contests.

Recruiting students is not a simple task, particularly if the students have had no prior experience with mathematics contests. The most efficient procedure might be to go into mathematics classes and invite the students, but for many reasons this may not be a viable approach. Regardless of how the recruiting is done, the objective of the coach will be to arrange a meeting for all students who are *potential* competitors. Once this initial meeting is esatablished, the coach's next task is to convince the students that they should participate in the contests. If the school is typical, these students will be the same ones who are involved in fifty other activities and are carrying-five academic subjects. What they do not need is another assignment. They will have some legitimate concerns. "Why should I join? Will it help me in college? Will it help me with my SAT? Will I get some relief from my work in my classes?" or more simply put, "What's in it for me?"

The coach needs to anticipate this when meeting with the students. There are several general guidelines for discussing the contests with the students:

• The primary purpose of the competition is to have fun with mathematics in the same sense that competitive basketball or baseball can be fun.

• The students will have the opportunity to tackle difficult problems they would not normally encounter in a regular class.

• Practice is important, but in contrast to other competitive activities, the students can practice on their own. They will be given practice problems, but they can do them on their own, they can work in groups, or they can work with the teachers. Flexibility is the key, because these students may be reluctant to participate in a program that will place a high demand on their time.

• Success or failure will have no permanent effect on the students' records—their grades will be neither raised nor lowered because of their performance—but what they learn from the contests may help them in all their future mathematics-related studies.

• The contest program can be paired with a mathematics club, which will give the students the opportunity to get together socially. Possibly the club can have an annual banquet or party. A coach should emphasize that the program is *extracurricular* and that every effort will be made to make it enjoyable.

If successful in recruiting students, a coach will probably also have to recruit other faculty to help with the program. They will be needed as judges, monitors, and officials during contests and as facilitators to help the students prepare for the contests.

The need for support personnel at a contest is crucial, and if there is no school-supported procedure for supplying faculty, it will be nearly impossible for the school to participate in the contests. Whenever faculty will be needed, there should be procedures similar to those used to supply faculty for other extracurricular events, such as basketball games or track meets.

Working with the students to prepare them for the contest is a different type of task. A coach can carry out this responsibility alone, but as stated earlier, there are benefits to the students and the faculty if more faculty are involved. If faculty other than the coaches are to work with the students, the following guidelines can help determine what their roles will be:

- The coach should be responsible for preparing and distributing all practice problems, including answers and other relevant information.

- The coach should be responsible for determining which students will participate in which categories, and the coach should make the final decision as to who will participate in the contests.

- Faculty should volunteer to work with specific topics or categories, and they should not be expected to work with more than a few students at any one time. If there are several contests throughout the year, a faculty member might work with several groups.

- The faculty should arrange to meet with their groups periodically, but the students should have time allocated when they can work by themselves. The students should not expect the teachers to "teach" the material, but rather they should consider the faculty to be resource people who can help them.

- The faculty should be encouraged to come to practice with open minds, not with formal preparations. The students need to develop a sense of independence in problem solving, and this occurs best when they are *not* presented with smooth, well-developed solutions to the problems.

- If there are special formats such as relays, two-person teams, or eight-person teams, the students should practice as teams. For such events the teachers should work with the students on teamwork.

- A final note of caution: Only faculty who volunteer should be working with the students, and none of the faculty should be given an excessive workload. It would be better for students to do the practicing by themselves than to have a reluctant teacher as their leader. To give an example of how the staff may be used to prepare the students for a contest, here is a possible distribution of responsibilities if there are five teachers and one coach working with the students:

> Freshman written competition (6–8 students): Teacher A
> Sophomore written competition (6–8 students): Teacher B
> Junior written competition (6–8 students): Teacher C
> Senior written competition (6–8 students): Teacher D
> Freshman/sophomore and junior/senior relay competition (4 teams):
>   Teacher E

Freshman/sophomore and junior/senior two-person team competition (all teams): Coach

Oral competitions (2 contestants): Coach

Freshman/sophomore and junior/senior eight-person team competitions: Independent practice

Calculator teams (6 students per team): Independent practice

The five teachers might be expected to work with their groups only six to eight times during the year, but the students will also be able to work as independent groups. As the school becomes more deeply involved in the contests, it is not unusual for teachers to offer to spend more time working with their groups.

In addition to recruiting students and coordinating the program, the coach has at least one additional task, which may be the most important one. That is to distribute resource materials for the students to use in preparing for the contests. A coach has to become an almanac of information about mathematics contests.

First, the coach should acquire an extensive collection of problems. The problems included in this book can serve as a starting set, but a coach should be alert to many other sources. Numerous books are available with many interesting problems, but a majority of these problems may be inappropriate for a mathematics contest. Such problems are better suited for other enrichment activities, which can be discussed in regular mathematics classes as individual project material. The best source of material would, of course, be questions from previous contests. But whatever sources are used, a coach has to collect as many questions as possible.

Once the questions are collected, another major task is organizing the material in a systematic way. For example, sets of relay questions would be considerably different from the sets of questions used for eight-person team competitions. Furthermore, any relevant information regarding techniques for competing in different categories should also be organized systematically.

Teachers who are not the official coaches but who are working with groups of students should not be expected to organize their own materials. For example, students who are preparing for the freshman written competition, and the teacher working with them, should be given sets of rules for the competition, the criteria that will be used to select the competitors, and possibly 100 problems with hints and solutions. Armed with these materials, the students and the teacher could then put together their own practice schedule.

The sample questions and formats that follow can serve as an initial source of material for a school that plans to enter into a mathematics contest program.

Competition Formats
and Sample Questions

Written Competition:
Open Categories, with Unweighted Questions

This competition is usually divided into four levels: Algebra, open to freshmen; Geometry, open to freshmen or sophomores; Advanced Algebra, open to freshmen, sophomores, or juniors; and Precalculus, open to all students. Each level has the same number of questions, usually ten or twenty, and a fixed time limit for the competition. A team usually consists of four to six students in each category, but a contest can easily be expanded for any number of students.

Each question has the same value, so the questions need not be sequenced. There is no partial credit for answers. The scores are recorded in a place on the examination or on an answer sheet. Any restrictions on the form of the answers must be included in the problem, or standard forms must be determined prior to the contest. For example, units such as "centimeters" or "kilometers per hour" must be either required of all contestants or omitted altogether. To make the contest interesting, the graders must be able to score a paper quickly; the time spent making decisions about answers detracts from the excitement of the contest. Some answers, however, may need interpretation, and that is why there should be a protest committee. For example, is the decimal notation $0.\overline{33}$ equivalent to the rational expression 1/3?

ALGEBRA

1. Simplify the following: $3 \times 2 - 12 \div 2 \times 3 = ?$

2. Find the largest prime integer between 210 and 220.

3. One class of 30 students averaged 52 on an exam. Another class of 25 students averaged 30 on the same test. What was the average of all 55 students?

4. If $5 \le a \le 10$ and $20 \le b \le 30$, then what is the maximum value of a/b.

5. If $x = -2$, then find x^{-x}.

9

6. A father is four times as old as his son. In 14 years he will be two times as old. How old will the father be when his son is 15?

7. What is the smallest whole number that 18 can be multiplied by to give a perfect cube?

8. If $-3 < x < 4$ and $1/2 < y < 3$, then $a < x/y < b$. What is $(a + b)$?

9. The product of the ages of three teenagers is 4590. How old is the oldest?

10. Simplify the following: $5.327^2 - 4.327^2 + 1 = ?$

11. Two pipes can be used to fill a swimming pool. The first can fill the pool in three hours, and the second can fill the pool in four hours. There is also a drain that can empty the pool in six hours. Both pipes were being used to fill the pool. After an hour, a careless maintenance man (obviously not a mathlete) accidentally opened the drain. In all, how long will it take for the pool to fill?

12. If $-8 < x < 2$, then $a \le |2 - |2 + x|| < b$. Find $(a + b)$.

13. Solve: $(3x^2 + 7x + 2)(x - 3) = (3x^2 - 8x - 3)(x + 2)$

14. Solve for x in terms of m and n:

$$\frac{2(x - m)}{3x - n} = \frac{2x + m}{3(x - n)}$$

15. Find the slopes of all lines through $(1,3)$ where the slope of the line is $-1/2$ of the value of the x-intercept.

16. The mean of four numbers is K. If the number 40 is added to the set of numbers, then the new mean for the five numbers is 14. What is K?

17. The circumference of a circle of radius m is 120. Find the circumference of a circle of area $270m$. You may wish to use these formulas:

$$A = \pi r^2 \text{ and } C = 2\pi r$$

18. For what values of c will the roots of the following function be reciprocals?

$$p(x) = 7x^2 + 4x + c$$

19. The hands of a clock are directly opposite at 6:00. When (exactly) will they next be opposite? Give your answer in minutes and fractions of a minute after the hour.

20. Form a quadratic equation whose roots are the following:

$$1 + m + \sqrt{1 + m^2} \quad \text{and} \quad \frac{2m}{1 + m + \sqrt{1 + m^2}}$$

When you have determined the quadratic, rewrite it in the form

$$x^2 + bx + c = 0$$

and give $b + c$ as your answer to the problem.

(Answers on page 107)

GEOMETRY

1. True or false? It is possible for five parts (sides and angles) of one triangle to be congruent to five parts of another triangle, but for the triangles not to be congruent to each other.

2. A can contains three tennis balls. Find the ratio of the height of the can to the circumference of the can.

3. Find the hypotenuse of a right triangle having legs of 121 and 363.

4. The difference between consecutive angles of a parallelogram is 72°. What is the measure of the smallest angle in the parallelogram?

5. In the following figure $\overline{AB} \parallel \overline{CD}$, $\overline{AD} \perp \overline{BC}$, $AB = 8$, $BC = 20$, and $CD = 32$. Find AE.

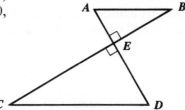

6. An equilateral triangle has an area of $9\sqrt{3}$. A point P is in the interior of the triangle and equidistant from each side. What is the distance from P to each of the sides?

7. The sum of the angles of a convex polygon is 1980°. How many sides does it have?

8. In the following figure find the area of the shaded region, given the following information. $\triangle ABC \sim \triangle EFG$, $\angle F$ is a right angle, $AB = 6$, $AC = 10$, $FG = 3$.

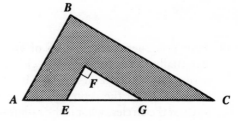

9. In the following figure $\overline{AD}$ bisects $\angle A$, and $\overline{DE}$ is parallel to $\overline{AC}$. $AB = 3$, $BD = 2$, and $DC = 3$. Find DE.

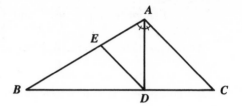

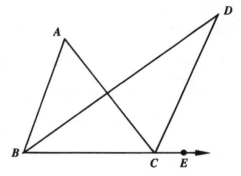

10. In the following figure, $\overline{BD}$ bisects $\angle B$, $\overline{DC}$ bisects $\angle ACE$, and $m\angle A = 60°$. Find $m\angle D$.

11. Find the area of the shaded region formed by constructing mutually tangent circles on the sides of a square whose sides have lengths of 4.

12. ABCD is a trapezoid with nonparallel sides $\overline{AB}$ and $\overline{CD}$. $AB = 8$ and $CD = 6$. If the angle bisectors of $\angle A$ and $\angle D$ intersect at a point E which is on BC, find the length of BC.

13. A 3-by-4-by-5 rectangular piece of cheese is cut in half by cutting diagonally along one face, forming two triangular prisms. Of the three possible such cuts, what is the minimum surface area for one of the triangular prisms?

14. Four congruent circles, each of which is tangent externally to two of the other circles, are circumscribed by a square of area 16. If a small circle is then placed in the center so that it is tangent to each of the circles, what is the radius of that circle?

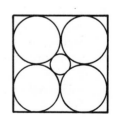

15. On a level plot of ground stand two vertical posts, one rising 36 feet, the  other 60 feet. From the top of each pole, a rope is stretched to the foot of the other pole. How far above the ground do the ropes cross?

16. Find the volume of a right pyramid having a square base if all eight edges have length 4.

17. In the following figure $ABCD$ is a square with sides of length 8. $DE = 6$, and $\overline{PQ}$ is the perpendicular bisector of $\overline{AE}$. Find the area of $PDEM$.

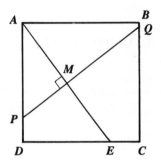

18. In the figure shown, $\triangle ABC$ is an isosceles triangle, M is the midpoint of $\overline{AC}$, $\overline{MG} \perp \overline{AB}$, $\overline{MH} \perp \overline{BC}$, $AC = 4$, and the area of $\triangle ABC = 8$. Find the area of quadrilateral $MHBG$.

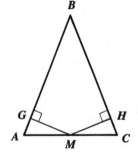

19. In the figure shown, $\overline{AB} \perp \overline{BC}$, $\overline{AE} \perp \overline{EC}$, D bisects $\overline{AC}$, and $\overline{DB} = 12$. Find the lengths of $\overline{DE}$ and $\overline{AC}$. Give your answer as the sum of DE added to AC.

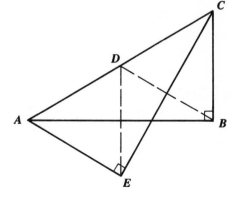

20. In the figure shown, *ABCD* is a rhombus with sides of length 4, *PQRS* is a square, and *BPQ* is an equilateral triangle. Find the area of the shaded region.

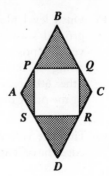

(Answers on page 107)

ADVANCED ALGEBRA

1. The first number is 13 larger than the second number, and the third number is 9 less than the first. The sum of the three numbers is 83. What is the smallest of the three?

2. If $f(x) = \dfrac{x}{2x - 1}$, find $f(3) - f(1/3)$.

3. The width of a room is two-thirds of the length. If 3 feet were added to the width and 3 feet were subtracted from the length, the room would be square. What is the length of the room?

4. From 1971 to 1980 the combined population of Minneapolis and St. Paul increased by 16%. St. Paul increased by 12% and Minneapolis increased by 18%. If St. Paul had a population in 1980 of 540 000, what was the population of Minneapolis in 1980?

5. Solve: $\sqrt{x^2 - 7} = \sqrt{x - 1}$

6. If $a \neq 0$, and $b \neq 0$, then solve the following system for (x, y) in terms of a and b:

$$\frac{x}{a} + \frac{y}{b} = 1$$

$$\frac{x}{2a} + \frac{y}{3b} = \frac{1}{3}$$

7. Find the least common multiple of the following polynomials, and for your answer give the coefficient of the x^3 term:

$$3x^2 + x - 2$$
$$3x - 8x + 4$$
$$x^3 - 2x^2 - x + 2$$

8. Simplify the following: $\dfrac{\left(2 - \dfrac{3}{x}\right)^2}{\left(4 - \dfrac{9}{x^2}\right)}$

9. Consider the arithmetic sequence 3, 7, 11, . . . What is the product of the 25th term and the kth term? Write your answer as a polynomial expression in k.

10. If $5^{2x} = 12 - 5^x$ and $5^{x+2} = M$, what are the possible values of M?

11. The roots of the function $f(x) = ax^2 + bx + 1$ are -2 and 3. What are the roots of $g(x) = bx^2 + ax + 1$?

12. If $\log_{10}3 = 0.48$ and $\log_{10}2 = 0.30$, how many digits are there in the expansion of 72^{10}?

13. A cubic equation of the form $ax^3 + bx^2 + cx + d = 0$, with a, b, c, and d being relatively prime integers, has solutions of 1/2 and $\sqrt{3} - 2$. What is b?

14. Consider all lines through $(-3, 2)$ such that the sum of the x and y intercepts equals twice the slope. What is the maximum slope of all such lines?

15. If $\dfrac{2}{x} - x = 2$, then find $\dfrac{4}{x^2} + x^2$.

16. Solve for x where n is an integer: $x^{2/5} + 6 = 5x^{1/n}$

17. A boat is rowed by 8 men, 4 on each side of the boat. Two of the men can only row on the right side and one man can only row on the left side. How many ways can the 8 men be placed in the 8 positions?

18. If $\dfrac{\dbinom{2n}{3}}{\dbinom{n}{2}} = \dfrac{44}{3}$, what is the value of n?

   (The notation means a combination of $2n$ things taken 3 at a time, etc.)

19. If Al and Bob play gin rummy, the probability that Al will win is 3/5. If they play five hands, what is the probability that Bob will win more hands than Al?

20. The Bulls and the Hustle are to play a best-of-five tournament—that is, the first team to win three games is the winner of the tournament. The Bulls are considered to be 2 to 1 favorites for any single game. In

the first game the Hustle wins, 83–81.What is the probability that the Bulls will still win the tournament?

(Answers on page 107)

PRECALCULUS

1. Solve for x: $3^{x^2} = 9^{2x+6}$

2. If $f(x) = x^2 - 2$ and $g(x) = \dfrac{1}{x^2 - 2}$, find $f[g(2)]$.

3. If the following is the graph of the cubic function $y = ax^3 + bx^2 + cx + d$, find $(a + b + c + d)$.

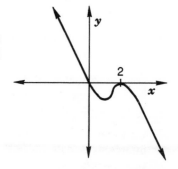

4. Find the equation of the line through the point $(4, -2)$ that is perpendicular to the line $2x - 5y = 53$. Write the equation of the line in the form $ax + by = c$, and then give the sum $a + b + c$ as your answer.

5. If $\tan \alpha = 2/3$, find $\sin 2\alpha$.

6. Find m: $\log_4 m = -5/2$

7. Solve for x: $\dfrac{|x - 2| - |x + 2|}{x} > 0$

8. Simplify: $\dfrac{[i^5(i^3 + i^2) - (i^6 - i^9)]^2}{i}$

9. Find the difference in the following repeating decimal real numbers. Give your answer as a reduced rational fraction.

$$(0.636363 \ldots) - (0.327327 \ldots)$$

10. Consider the conic equation $x^2 - y^2 = 4$. What is the equation of the conic that is determined if the axes are rotated $45°$ in a clockwise direction?

11. If $f(x) = x^2 - 7x + 11$ has roots r and s, find $r^2 + s^2$.

12. Let $f(x) = 3x + 2$ and $g(x) = 2x^2$. Find the largest value of x for which $f(g(x)) = g(f(x))$.

13. Find two positive integers whose arithmetic mean is one-half more than their geometric mean and where the larger integer is 5 more than the smaller integer. Write the larger as your answer.

14. Find the coefficient of the x^3 term in the expansion of $(1 + 3x + 2x^2)^6$.

15. If $\sin \alpha = 0.64$, find $8\left(\cos \dfrac{\alpha}{2}\right)\left(\cos \dfrac{\alpha}{4}\right)\left(\cos \dfrac{\alpha}{8}\right)\left(\sin \dfrac{\alpha}{8}\right)$.

16. Assume θ and ϕ are both in the first quadrant. If $\tan \theta = 3/4$ and $\sec \phi = 3/2$, find $\sin (\theta + \phi)$.

17. Consider a circle of radius 4 with center $(0, 0, 0)$ that is in the plane $x - y = 0$. What is the equation of the projection of this circle into the y-z plane?

18. Determine the equation of an ellipse whose major axis is twice the length of the minor axis, whose center is at the origin, and that contains the points $(1, 2)$ and $(\sqrt{5}, \sqrt{3})$. Write your answer in the form $ax^2 + by^2 = c$, where a, b, and c are relatively prime integers.

19. Find the value of the following infinite continued fraction:

$$\cfrac{1}{1 + \cfrac{1}{2 + \cfrac{1}{1 + \cfrac{1}{2 + \cfrac{1}{1 + \cfrac{1}{2 + \cdots}}}}}}$$

20. A fourth-order determinant has as elements in the ith row and the jth column the binomial coefficient

$$\binom{i + j}{j}.$$

Find the value of this determinant.

(Answers on page 108)

Written Competition:
Topics Identified, with Weighted Questions

This category usually has four levels of competition, but in contrast to the open competition, the topics are identified and all the questions relate to that topic. For example, if the topic were "Algebra: Systems of Equations," then the questions would be related to this topic and weighted by difficulty. If there are five questions progressively worth 1, 2, 3, 4, or 5 points, a contestant answering all five questions correctly would receive 15 points.

An advantage to this approach is that a student's effort in preparing can be a more significant factor than in an open competition. If your goal is to encourage students to study for the competition, this format is better than the open competition.

Although the number of students taking the examination can be exactly the same as in the open competition, this format offers an opportunity for greater participation. Because there are fewer questions per topic than in an open category, the contestants need less time to complete the examination. For example, two topics can be tested for thirty minutes each rather than a single examination taking sixty minutes. There can even be three topics at each level.

There can also be greater flexibility. Although a team may consist of as few as four contestants, it is still possible for a school to have eight or twelve students participating at each level, if some students enter for only one topic. Yet, a smaller school can use the same four contestants for all topics at their level.

Writing the questions is difficult. The writer may need to know some restrictions for questions in each category. The order of difficulty of the questions can only be estimated, and the question writer can be far off in his or her estimate. As with the other written competition, questions may arise, and there should be a protest committee available.

The examples that follow are taken from actual contests, and the order of difficulty proved to be correct, at least for the students in those contests. There are six topics at each level.

FRESHMAN LEVEL

Factoring

1. $3x^4 - 48$

2. $6x^2y - 15y - 5x + 18xy^2$

3. $50x^2 - 401x + 804$

4. $m^2(m - 4) - 2m^2 + 5m + 12$

5. Find the values of m for which the following polynomial will be factorable into the product of two linear factors:

$$2x^2 + mxy - 6y^2 + 5x + 9y - 3$$

Factors and Primes

1. Suppose $N = (2^5)(3^9)(5^7)$ and $M = (2^7)(3^2)(5^4)$. Find the greatest common divisor of M and N. Give your answer as an integer.

2. The least common multiple of two numbers is 60 and the greatest common divisor is 6. What are the two numbers?

3. If $M = 225$ and $N = 444$, determine the product of the greatest common divisor of M and N and the least common multiple of M and N. Write your answer as an integer.

4. Let $M = 3645$ and $N = 4368$. Find the least perfect square that is a common multiple of M and N. Leave your answer in the form $a^m b^n c^p$ ..., where $a < b < c < \ldots$.

5. Tom said, "I'm in the *prime* of my life, but I must admit that two years ago I was also having a prime year." Judy said, "I know that six years ago you were just an odd square." We know that Tom can vote but is certainly not in the *Guinness Book of World Records*. How old is Tom?

Logic Puzzles

1. Which of the following statements is true?
    a) Exactly one of these statements is false.
    b) Exactly two of these statements are false.
    c) Exactly three of these statements are false.
    d) Exactly four of these statements are false.
    e) Exactly five of these statements are false.

2. One hundred typical high school freshmen were interviewed about their reading habits. Sixty-three said that they read *Mad Magazine*, and forty-one said that they read *The Saturday Review*. Ten said that they read both magazines. How many read neither?

3. Two Holstein and three Guernsey cows give as much milk in three days as two Guernsey and four Holstein give in two days. Which type of cow gives more milk per day?

4. When each letter in the following problem is replaced by an appropriate digit, the solution to the problem is correct. The O, T, F, and E cannot be zero, and each letter must represent a different digit. What digit is represented by F?

$$
\begin{array}{r}
O\,N\,E \\
T\,W\,O \\
\underline{F\,I\,V\,E} \\
E\,I\,G\,H\,T
\end{array}
$$

5. Four men—Al, Bob, Carl, and Dan—were speaking of their wives. They were not well acquainted and the statements they made were not all accurate. In fact the only thing that is certain is that when a man mentions his wife, his statement is correct.

Al: Adell is Betty's mother.
    I have never met Diane.
Carl: Diane is Al's wife.
    Adell is Betty's older sister.

Bob: Carl's wife is either Adell or Diane.
Betty is the oldest.
Dan: Carol is my daughter.
    Adell is older than my wife.

Who is Al's wife?

Modular Arithmetic

For any problem modulo K, use the integers $\{0, 1, 2, \ldots, K - 1\}$

1. Determine the value of N in the set $\{0, 1, 2, 3, 4, 5, 6\}$ such that
$$53\,814 \equiv N(\text{modulo } 7).$$

2. Find all solutions modulo 8 for which
$$X^2 \equiv 0.$$

3. Simplify the following:
$$4^{55} \equiv \underline{\hspace{3cm}}(\text{modulo } 7)$$

4. If 1 January 1957 was a Tuesday, what day of the week was 1 February 1960?

5. Determine the value of N such that
$$(24\,737)^2 \equiv N(\text{modulo } 24).$$

Number Bases

1. What is the base 10 equivalent of the largest 4-digit integer written in base 5?

2. For what base is $5 \times 24 = 212$?

3. Insert the missing term:

$$10, 11, 12, 13, 14, 15, 20, 22, 30, \underline{}, 1100$$

4. Change the following number, which is written as a base 12 (duodecimal) number, into its equivalent reduced rational fraction written in base 10:

$$\frac{149}{200}$$

5. The letters a, b, c, and d represent four digits. If the following statements are true, determine the values of a, b, c, and d.

In base 7: $\quad a\,b$ In base 9: $\quad a\,b$ In base 8: $\quad c\,d$
$\underline{+c\,d}$ $\qquad\qquad \underline{+c\,d}$ $\qquad\qquad \underline{-a\,d}$
$\quad\;\;132$ $\qquad\qquad\quad\; 110$ $\qquad\qquad\quad\;\; 27$

Rational Arithmetic Using Calculators

1. Which of the following rational numbers is the largest?

$$\frac{17}{37}, \quad \frac{1738}{3793}, \quad \frac{173765}{379258}$$

2. Simplify the following:

$$251.3 + \left[\frac{53.63 - 5.79}{6.57 \times .328}\right] \div 36.5 + 67.38$$

3. The volume of a sphere is $(4\pi r^3)/3$. What is the radius of a sphere having a volume of 1000? Assume $\pi = 3.14159$.

4. Determine the following to three decimal places:

$$\cfrac{3}{1 + \cfrac{3}{1 + \cfrac{3}{1 + \cfrac{3}{1 + \cfrac{3}{\cdots}}}}}$$

5. Tom wants to double his investment in exactly seven years. What percentage annual interest will allow him to do this? (Remember that the interest from one year is added to the principal to determine the interest for the next year.) Give your answer to one decimal place.

Word Problems

1. A brick weighs 5/8 of a kilogram plus 5/8 of its weight. How much does the brick weigh?

2. In our continuing support for the work ethic, we have another earnings problem. Al and Brian each earned $10 per day. Because of hard work and good grades in mathematics, Brian was given a $2 per day raise. In a 60-day period, sometime during which Brian was given the raise, the two earned a total of $1244. How many days had Brian worked at the lower rate before he was given the raise?

3. If I ride 2 km to school at 4 km per hour and ride home at 10 km per hour, what is my average speed for the two trips?

4. Mr. Smith is a compulsive gardener. He has to arrange his trees into square arrays. After forming a large square, he finds 46 trees are still unplanted. Therefore he goes out and buys 13 more trees and forms a new square array one row larger than the previous square. How many trees has he planted?

5. A train passes a pole in thirty seconds, and it passes completely through a 500-meter tunnel in one minute and twenty seconds. How long is the train?

Sets and Venn Diagrams

1. True or false? If the statement "All cats are dogs" is not true, then the statement "All cats are not dogs" is true.

2. How many subsets are there for the following set: {a, e, i, o, u}?

3. An environmental group made a survey of 1000 people in different age groups to determine the environmental problems that concerned them. The following table gives the results. The letters in parentheses designate the sets.

	Water Pollution (W)	Air Pollution (A)	Over-population (P)	Energy Conservation (E)
Under 25 (U)	94	62	110	34
25–40 (M)	208	70	25	81
Over 40 (O)	31	180	16	59

Find the number of people in (A ∪ E) ∩ M'. (*Note:* The symbol ' means complement over the universal set.)

4. A recent survey of 200 mathletes provided the following information:

    95 are taking chemistry.

    115 are taking a foreign language.

    55 are taking creative transactional analysis (CTA).

    30 are taking chemistry and a foreign language.

    25 are taking chemistry and CTA.

    35 are taking a foreign language and CTA.

    10 are taking chemistry, a foreign language, and CTA.

How many mathletes are not taking any of the courses listed?

5. Ruralburg has one movie house and a potential viewing population of 2800. One week 1800 people saw *Jaws II*, the next week 1640 people saw *The Revenge of the Pink Panther*, and one week later 2210 people saw *Star Wars*. Assuming no person saw the same movie twice, what is the minimum number of persons who would have seen all three movies?

(Answers on page 108)

SOPHOMORE LEVEL

Arithmetic and Geometric Progressions

1. Find the sum of the first 99 positive integers that are divisible by 7.

2. Find two numbers whose sum is 20, and whose arithmetic mean is 2 more than the geometric mean.

3. A student has a 4.2 average for 10 credits. What must be his average for 5 additional credits to improve his average to 4.40?

4. Suppose that each swing of a pendulum bob is 94% of the length of the preceding swing. If the first complete swing is 0.84 meters, what is the total distance that the bob travels before coming to rest?

5. A farmer is told that the amount of algae in his pond is doubling each day and that the pond will be completely covered after 30 days. What portion (fractional part) of the pond will still be clear after 20 days?

Circles

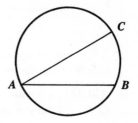

1. In the following circle $m\widehat{ABC} = 230$ and $m\widehat{ACB} = 190$. Find $m\angle CAB$.

2. The following circles are concentric. The inner circle has a radius of 4 cm, and the area of the region
   between the circles is 20π cm². What is the diameter of the larger circle?

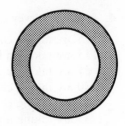

3. Two 10-cm chocolate circles are
   packed next to each other in a rectangular box as shown here. What
   is the area of the box that is *not*
   covered with chocolate?

4. In the following figure, $m\widehat{AB}$
   $= 110$, $m\widehat{ED} = 80$, $m\angle P =$
   15, and the $m\angle AOE = 55$.
   Find $m\widehat{CB}$.

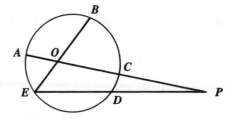

5. In the following figure, $\overline{EC}$ is a
   tangent, and $\overline{AB}$ is a diameter
   to the circle. If $CD = 6$, $DA =$
   10, and $AF = 8$, find FE.

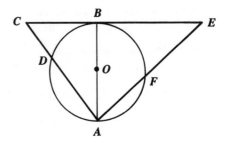

Coordinate Geometry

1. For what value of a are the following points collinear?
$$P(3, -1), Q(7, 1), \text{ and } S(2635, a)$$

2. A line has an x-intercept of $(8, 0)$ and contains the point $(4, 1)$. Find the
   length of the segment having the x- and y-intercepts as its endpoints.

3. $ABCD$ is a parallelogram with coordinates $A(-3, -2)$, $B(1, -5)$, and
   $C(9, 1)$. Find the sum of the coordinates of the point D.

4. $\triangle ABC$ has coordinates $A(5, 5)$, $B(3, -1)$, and $C(7, 3)$. Find the coordinates of the reflection of A over the line $\overleftrightarrow{BC}$.

5. $\triangle ABC$ has coordinates $A(0, 0)$, $B(8, 0)$, and $C(3, 8)$. Find the coordinates of the intersection of the altitudes.

Geometry of the Right Triangle

1. Find the length of the diagonal of a rectangular solid having sides of lengths 3, 4, and 12.

2. In the right triangle ABC, if D is a point on the hypotenuse, $\overline{AC}$, such that $AD = DC = BC$, then what is the measure of $\angle ABD$?

3. In the following plane figure, AB, EF, and DC are each perpendicular to BC. $AB = 3$, $DC = 6$, and $BC = 15$. Find EF.

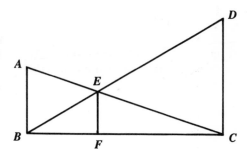

4. In the following plane figure, $\angle C$ is a right angle and $\overline{DM}$ is the perpendicular bisector of $\overline{AB}$. If $AC = 6$ and $BC = 10$, then find the area of $\triangle MBD$.

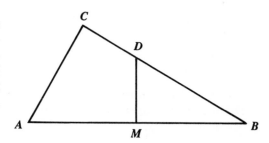

5. Find the area of a right triangle having a perimeter of 20 that has an altitude to the hypotenuse of length 5.

Polygons and Polyhedra

1. The following quotation is taken from *Flatland*, a novel originally published in 1884 about a two-dimensional world inhabited by geometric figures:

Our middle class consists of equilateral triangles. Our professional men are squares and pentagons. Next above these come the nobility, of whom there are several degrees, beginning at hexagons and from thence rising in the numbers of their sides. When the number of the sides becomes so numerous and the sides themselves so small that the figure cannot be distinguished from a circle, he is included in the circular or priestly order; and this is the highest class of all. [Abbott, Edwin A. *Flatland*. New York: Dover Books, 1952, pp. 8–9.]

If every resident has the same perimeter, who has the largest radius, that is, the radius of the circumscribed circle?

2. For the regular five-pointed star shown, find the measure of a vertex such as ∢*A*.

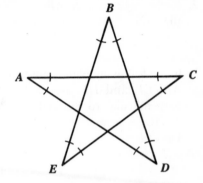

3. A square is inscribed in a regular octagon using segments that join every other vertex of the octagon. If the length of the square is 4, find the area of the octagon.

4. How many vertices does a dodecahedron have? (Hint: For all polyhedra, $V + F - E = 2$.)

5. A cube is 4 cm on each edge. The segments connecting the centers of each *face* of the cube form an octahedron. What is the volume of that octahedron?

Perimeters and Areas

1. True or false? For any two similar triangles, if one pair of corresponding sides are in a ratio of 3 to 1, then the areas are in a ratio of 9 to 1.

2. Trapezoid *ABCD* has parallel sides $AB = 6$ and $CD = 12$, and it has an area of 72. *E* is a point on $\overline{CD}$ such that $\overline{BE}$ is parallel to $\overline{AD}$. Find the area of △*BEC*.

3. △*ABC* is a right triangle with an altitude to the hypotenuse $\overline{AD}$. If *BD* = 4 and *BC* = 3, what is the area of the triangle?

4. Given an equilateral triangle with sides of length 4, let P be any point in the interior. Find the sum of the three perpendiculars to the sides from the point P.

5. In the following figure, P and Q are points of trisection of sides $\overline{HD}$ and $\overline{AR}$, respectively. $HARD$ is a rhombus and M is the midpoint of $\overline{PQ}$. If the area of $\triangle HMP$ is 4, find the area of the entire rhombus.

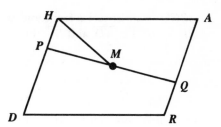

Systems of Equations

1. Are the following equations consistent?
$$3x - 5y = 2$$
$$x + 2y = 8$$
$$\frac{5}{8}x + \frac{3}{4}y = 3$$

2. Solve for (x, y):
$$14y = 46x - 62$$
$$7y = 104 - 22x$$

3. Solve for (a, b, c, d):
$$a + b + c = 0$$
$$b + c + d = 1$$
$$a + c + d = 2$$
$$a + b + d = 3$$

4. Solve for (x, y):
$$y^2 = (x - 37)^2 - 60$$
$$x - y = 27$$

5. The following system of equations has solutions at the points $(2, 3)$ and $(-5, -4)$. Find the value of a.
$$ax^2 + by^2 = e^2$$
$$6x + cy = e$$

Similar Triangles

1. $\triangle ABC \sim \triangle DEF$, with $\angle A \cong \angle D$, $\angle B \cong \angle E$, $AB = 14$, $AC = 15$, $BC = 16$, and $DF = 21$. Find the perimeter of $\triangle DEF$.

2. $ABCD$ is an isosceles trapezoid with $AD = BC = 15$, $AB = 30$, and $CD = 20$. Let $\overrightarrow{AD}$ and $\overrightarrow{BC}$ intersect at E. What is the length of DE?

3. In the following figure, $AB = BC$ and $AC = AD$. If $AC = 6$ and $CD = 4$, find BD.

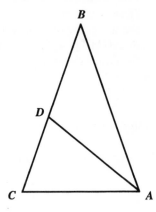

4. $\triangle ABC$ is a right triangle with $m\angle A = 90°$, and $\overline{AD}$ is the altitude to the hypotenuse. If the legs of the triangle are 2 and 4, what is the length of AD?

5. In the following figure, AB is a diameter, AC and BD are tangents to the circle, and AD and BC intersect at the point E on the circle. If $AB = 6$ and $AC = 5$, find BD.

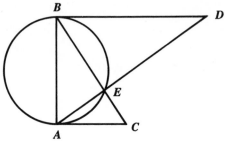

Word Problems

1. Twenty-five stamps, some costing 18¢ and some costing 22¢, cost a total of $4.90. How many were 18¢ stamps?

2. A and B are two integers. If A is divided by B, the quotient is 4 and the remainder is 5. If $2A$ is divided by B, the quotient is 9 and the remainder is 3. What is A?

3. The difference of two numbers is 3, and the difference of their squares is 27. Find all such pairs of numbers.

4. The combined ages of all the members of a family—a father, mother, son, and daughter—is 68. The father is 4 years older than the mother, and the daughter is 2 years older than the son. Four years ago the combined age was 53. How old is the father now?

5. Alpha, Beta, and Celia work for the local transportation company, counting paper dollars. To count one batch of paper dollars, Alpha would take 10 hours longer than the three of them working together. Beta would require only 4 hours more than the three of them working together. Celia, however, would require twice as long as the three of them working together. Who is the fastest worker, and how long would it take her to complete the task?

(Answers on page 109)

JUNIOR LEVEL

Complex Numbers

1. If $\sqrt{-1} = i$, the imaginary unit, then simplify the following, using the unit i:

$$\sqrt{-\frac{3}{2}} + \sqrt{-\frac{2}{3}}$$

2. Simplify the following, given v is a real number and $v > 0$:

$$\frac{\sqrt{-25v^8}}{\sqrt{-20v^6}}$$

3. Solve for x: $x^2 = 5 - 12i$

4. If two roots of the polynomial equation $x^6 - 6x^5 + 14x^4 - 22x^3 + 25x^2 + 8x = 60$ are $2i$ and $2 - i$, what are the real roots?

5. Let r_1 and r_2 be the roots of the equation $x^2 + bx + c = 0$. If $r_2 - r_1 = 6i$ and $r_2 \cdot (r_1) = 2$, what is b?

Conic Sections

1. Find the center of the ellipse $x^2 + 2y^2 - 2x - 4y = 9$.

2. Which of the following parabolas has the greatest minimum value? (Write "A," "B," or "C" as your answer.)

    (A) $f(x) = 3x^2 - 4$
    (B) $g(x) = x^2 - 10x + 20$
    (C) $h(x) = x^2 - 3$

3. Find the eccentricity for the following hyperbola:
$$45(x - 4)^2 - 36(y + 1)^2 = 80$$

4. Jim has a piece of wood that is 1 meter by 2 meters. He wants to make an elliptical table top having these two dimensions as the minor and major axes. He can draw the ellipse by following these steps: (1) pounding two small nails into the wood; (2) slipping a loop of string over the nails, allowing for some slack; (3) pulling the loop taut with a pencil; and (4) keeping the loop taut while drawing the ellipse. Question: How far from the edges, as shown in the figure below, should Jim put the nails? You may leave your answer in simplest radical form, or estimate to the nearest mm. Give your answer as (A, B).

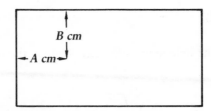

5. An isosceles triangle has its base on the y-axis with one base vertex at the origin. Find the equation of the locus of the nonbase vertices with a positive x-coordinate that generates isosceles triangles having a perimeter of 20. Write your answer as a function of x.

Inequalities

1. True or false? For all real values of x, $x^2 + 1 \geq 2x$.

2. Find all real values of x for which the following inequality is true:
$$|2x - 1| > 2$$

3. If x and y are positive real numbers such that $x + y = 10$, then find the minimum value of $\dfrac{1}{x} + \dfrac{1}{y}$.

4. Find all real values of x for which $2x^2 - x$ is less than 3.

5. Find all the real values of x for which the following inequality is true:
$(x^4 - 10x^2 + 9)(37x - 53)^2 < 0$.

Lines

1. If A, B, and C are collinear and $AB = 5\frac{3}{8}$, $AC = 6\frac{3}{4}$, and $BC = k$, such that $k > 10$, find the value of k.

2. Suppose A, B, C, and D are collinear and in that order such that $AB/BD = 2/3$ and $AC/CD = 3/4$. If $BC = 2$, then $AD = k$. Solve for k.

3. A line is given two Euclidean coordinate systems. (For example, it may be metric using centimeters, or it may be English using an inch system.) In one system the points are given the coordinates -3, 5, and 17 respectively. When using the second system the coordinates of P and Q are 5 and -2 respectively. What is the coordinate of R, using the second system?

4. $\triangle ABC$ has coordinates $A(0, 0)$, $B(3, 5)$ and $C(-1, 5)$. Find the point of intersection of the altitudes.

5. Find the equation of the line containing the points $(\sqrt{2}, \sqrt{8})$ and $(2, 0)$. Write the equation in the form $y = mx + b$ where m and b are real numbers in simplest radical form.

Similarity

1. In $\triangle ABC$, $AB = 4$, $BC = 6$, and $AC = 8$. In $\triangle PQR$, $PQ = 10$, $PR = 12$, and $QR = 14$. Is $\triangle ABC$ similar to $\triangle PQR$?

2. The following figure is a pyramid with a square base $BCDE$. The plane determined by $PQRS$ is parallel to the plane determined by $BCDE$. If the area of $BCDE$ is 50, $AP = 2$, and $PB = 3$, what is the area of $PQRS$?

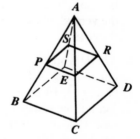

3. In the following figure, what values of x will make $\overline{BC}$ parallel to $\overline{DE}$?

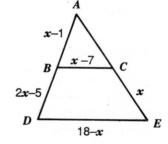

4. In the following figure, if $\overrightarrow{BE}$ bisects $\angle B$ and $\overrightarrow{CD}$ bisects $\angle C$, $AD = 3$, $AE = 4$, and $EC = 6$, what is the length of BC?

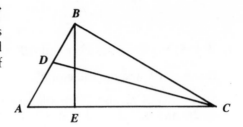

5. In the following figure, $\angle ADB$ ◊ $\angle C$, $AD = 5$, $AB = 4$, and $DE = 6$. If $\overline{AD}$ is parallel to $\overline{BE}$, find BD.

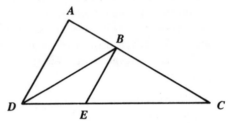

Surface Areas and Volumes

1. Find the volume of a cube with a surface area of 48.

2. What is the volume of the largest cylinder that can be made from a rectangular sheet of metal 6 cm by 8 cm if the circular bases are not made from the sheet of metal?

3. A right circular cone has a height of 10 cm and a volume of 144π cm^3. If a plane intersects the cone perpendicular to the altitude and 5 cm from the vertex, what is the volume of the smaller cone that is determined by the cut?

4. *ABCD* is an isosceles trapezoid with an upper base, $\overline{AB}$, 6 cm long, a lower base, $\overline{CD}$, 14 cm long, and a height of 3 cm. If the trapezoid is rotated around base $\overline{CD}$, it will determine a solid of revolution. What is the surface area of this solid?

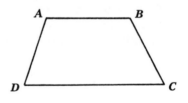

5. In the following trapezoid, $\overline{PQ}$ is parallel to $\overline{RS}$, $PQ = 4$, $QR = 10$, $RS = 25$, and $PS = 17$. If the trapezoid is rotated about its shorter base, $\overline{PQ}$, then what is the volume of the solid that is determined?

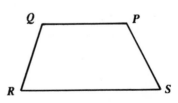

Probability

1. Which is more likely to occur in one trial? (Give "A" or "B" as your answer.)
   (A) Drawing an ace or a spade in one draw
   (B) Rolling a 2, 3, 7, or 11 with an ordinary pair of dice

2. What is the probability that if three cards are drawn from an ordinary deck of cards, no two will have the same value or will be the same suit?

3. Assume that the probability of Artis Gilmore making a free throw is ¾. He is fouled and given three attempts to make two points. (If he makes the first two shots, he does not try a third shot.) What is the probability that he will score two points?

4. Harold has six different textbooks on his shelf. What is the probability that if placed there randomly, his algebra book and his English book will not be adjacent to each other?

5. Al sells a set of encyclopedias for each five visits, while Bill is able to sell a set every four visits. If Al makes two visits and Bill makes three visits, what is the probability that Al will make more sales than Bill?

Word Problems

1. The difference of the squares of two consecutive positive integers is 121. What is the larger of the two?

2. Bill bought some eggs at four for 25¢. He ate one-fifth of them, then sold the rest at three for 25¢ and still made a profit of 25¢. How many eggs did he eat?

3. Al, Bob, and Chuck walk at 4 km per hour, 5 km per hour, and 6 km per hour respectively. Al starts walking along the Yellow Brick Road at 7:00 A.M. Bob starts along the same route at 9:00 A.M. At what time should Chuck leave so that the three meet at the same time?

4. A segment that is 6 cm long is divided into two parts. The area of a rectangle whose sides are the length of the whole segment and the shorter part is the same as the area of a square whose sides are the same length as the longer part. What is the length of the shorter part?

5. A group of laborers all worked at the same pace and received the same rate of pay per hour for the time that they worked. If all had worked together, they could have completed a job in 24 hours. But instead of beginning at the same time, they each started separately at equal inter-

vals. Then each continued until the job was completed. If the first worker was paid seven times as much as the last worker, how much time did it take to complete the job after the first worker began?

(Answers on page 109–110)

SENIOR LEVEL

Complex Numbers

1. True or false? If α, β are complex numbers, and $\bar{\alpha}$ means the conjugate of α, then $(\alpha + \beta)^2 = (\bar{\alpha} - \bar{\beta})^2$.

2. Evaluate the following: $\left(\dfrac{\sqrt{2} + i\sqrt{2}}{2}\right)^4$.

3. Solve for all complex solutions (you may express your answer in polar form): $z^3 = -32\sqrt{2} + 32i\sqrt{2}$.

4. Let $\bar{z}$ be the complex conjugate of z. If $z = 3 - 7i$, find $z^4 - \bar{z}^4$.

5. Solve for all complex solutions (you may express your answer in polar form): $z^2(3 - z^2) = 9$.

Coordinate Geometry

1. True or false? If $PQRS$ has coordinates $P(-2, -4)$, $Q(-10, 10)$, $R(14, 26)$, and $S(26, 12)$, then it is a rectangle.

2. Find the values of k so that the line containing the points $(2 - k, k - 5)$ and $(-2, -4)$ is parallel to the line containing the points $(k, k - 3)$ and $(1, -4)$.

3. Find the distance between the point $(2, -3)$ and its reflection in the line $y = 3x - 4$.

4. Find the area of the triangular region determined by the three intersecting lines:

$$3x + 4y = 11$$
$$4x - 3y = -27$$
$$7x + y = 9$$

5. Find the algebraic equation of the locus of points that determine triangles of area 40, with one side of the triangle being the segment determined by the points $(0, 0)$ and $(6, 8)$. Write the equations in the form $ax +$

$by = c$, where a, b, and c are integers and a, b, and c have no common factors other than 1 or -1.

Diophantine Equations

1. Find all positive integral solutions for the following equation (give the largest value of x as your answer): $13x + 7y = 112$

2. Solve the following system of equations for all positive integral solutions (give the largest value of x as your answer):

$$3x + 2y + 5z = 88$$
$$2x - 5y + 3z = -2$$

3. Find the hypotenuse of length greater than 50 but less than 75 for which the hypotenuse and one of the legs are consecutive integers.

4. A man orders sixteen pieces of wood totaling 155 feet. Eight of the pieces are of the same length, and each of the other eight are either 2 feet or 3 feet shorter than the eight of the same length. (Each piece is an integral number of feet.) How long is each of the eight longest pieces?

5. Find the area of a triangle that has four consecutive integers for the altitude and the three sides, the least being the altitude. (You may wish to use Hero's formula for the area: $A = \sqrt{s(s - a)(s - b)(s - c)}$, where a, b, and c are the sides and s is half of the perimeter.)

Logs and Exponents

1. If $f(x) = 25\dfrac{1}{x}$, find $f(-2)$.

2. If $\log 2 = 0.301$ and $\log 3 = 0.477$, find $\log 14.4$.

3. Solve for x: $\log (x) > \log 2 + \dfrac{1}{2} \log (x)$

4. Solve for x: $\log_x (27) = \log_4 (3)$

5. If $x > 0$, $y > 0$, $x^2 + y^2 = 33$, and $\log_2(x) + \log_2(y) = 3$, find $|x + y|$.

Probability

1. The integers 1 through 300 are written on identical pieces of paper, and one of the sheets is drawn randomly. What is the probability that the

number drawn is divisible by 3, or that the sum of the digits will be divisible by 3?

2. If seven fair coins are tossed, what is the probability that at least four will be "heads"?

3. Consider the quadratic polynomial $x^2 + ax + b$. If a and b are randomly selected to be positive integers that are less than or equal to 10, what is the probability that the polynomial is factorable over the integers?

4. A bag contains three red balls and two blue balls. Tom draws one ball in each hand. He looks at one hand and observes the color. What is the probability that the ball in his other hand is a different color?

5. A bag contains three red balls and three blue balls. Jim draws three balls at random and then replaces them with green balls. If he then draws three more balls at random, what is the probability that the three will be of different colors?

Theory of Equations

1. How many imaginary roots does the following polynomial equation have?

$$x^8 + 5x^3 + 3x - 9 = 0$$

2. Find the remainder when the following polynomial function is divided by $x + 1$:

$$p(x) = 37x^{50} - 3x^{35} + 2x^{17} - 21x^{10} + x^5 - 5$$

3. Given the equation $x^3 + 9x^2 + bx - 64 = 0$, where it is known that the roots are all positive and form a geometric progression, find the value of b.

4. The following is the graph of a cubic function. When written in the form $p(x) = ax^3 + bx^2 + cx + d$, with a, b, c, and d being integers, what is the value of c?

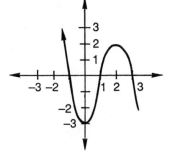

5. Suppose the equation $x^3 - 3x^2 + 7x - 4 = 0$ has roots r, s, and t. Find $\dfrac{1}{r^2} + \dfrac{1}{s^2} + \dfrac{1}{t^2}$.

Trigonometry

1. Solve for x in the interval $0 \leq x \leq \pi/2$:
$$4(1 - \sin^2 x)(\sec^2 x - 1) = 1$$

2. Find an equivalent simplified form of the following trigonometric function. (Your answer should be of the form $f(\theta)$, where $f(\theta)$ is one of the six elementary functions.)
$$\frac{\sin (2\theta)}{1 + \cos (2\theta)}$$

3. Allowing for slight visual error, the following is a general graph of a function of the form $y = a + b \sin c(x + d)$. Find a, b, and c. Write your answer as an ordered triple (a, b, c).

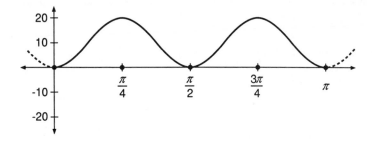

4. A spring acts with a simple harmonic motion. The amplitude, or maximum displacement, is 9 meters and the period of oscillation is 12 seconds. If at $t = 0$ the displacement is -9 meters (that is, the spring is pulled down and released), what is the displacement at $t = 14$ seconds?

5. A right pyramid has a square base with sides $2\sqrt{2}$ cm in length, and it has a height of 6 cm. What is the cosine of one of the upper vertex angles of the triangular faces?

Word Problems

1. The sum of a number and its reciprocal is 25/12. If the number is greater than 1, what is the number?

2. Harry Smoothvoice is paid $120 for a radio broadcast. If the broadcast had lasted 5 minutes longer, his rate of pay per hour would have been $120 less. What was the length in minutes of his broadcast?

3. Albergiliquious, Blesexillia, and Charchonetelly (hereafter referred to as A, B, and C) painted a fence. If A had painted alone, it would have taken

him six hours more than the three working together, but B would have needed fifteen hours more. C would have required twice the time that it took for the three of them working together. How long would it have taken A working alone?

4. One man starts riding a bicycle at 2.5 miles per hour. Another starts from the same point at the same time at 6 miles per hour, heading in a direction perpendicular to the direction of the first man. After what length of time will they be 26 miles apart?

5. Argyle the ant has his home at the vertex of a 3 × 4 × 5 rectangular *solid*. He decides to take a trip to the vertex that is diagonally opposite his home. Argyle is of course a brilliant but lazy ant and wishes to take the shortest possible route from his home to the other vertex. He therefore charts the route and calculates the distance. If Argyle can walk anywhere *on* the solid, how far will he travel? (Hint: It might help to unfold the solid.)

(Answers on page 110–11)

Eight-Person Team Competition

There can be many variations with this format. The team may consist of two students from each of the four grade levels; there may be two levels of competition, freshman-sophomore and junior-senior; there may be unbalanced teams, such as three seniors and two juniors, and two sophomores and one freshman; or there may be teams of six or ten persons. The size and structure should be determined by the number of students the competing schools would like to have participate.

An eight-person team competition usually consists of a twenty-question examination with a twenty-minute time limit. The members of the team can divide their tasks in any manner they choose, and at the end of the twenty minutes the team score is based on the total number of questions answered correctly, typically at 5 points per question.

The questions in this competition are generally the broader, more insightful types. Therefore, group techniques such as brainstorming or finding alternative strategies are often useful. The first group of examples that follow are open to four grade levels; the other two groups are restricted to two grade levels.

FRESHMAN–SENIOR LEVEL

1. A pentomino is a polygon made up of five congruent squares. Each square shares complete sides with all other squares it touches. Two pentominos are equivalent if they have the same shape, although you might have to rotate or reflect them to show they are equivalent.

   Examples:

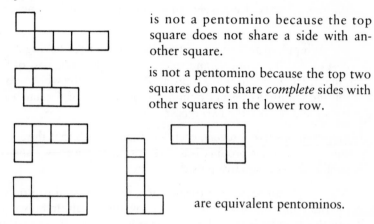

   is not a pentomino because the top square does not share a side with another square.

   is not a pentomino because the top two squares do not share *complete* sides with other squares in the lower row.

   are equivalent pentominos.

   How many different nonequivalent pentominos are there?

2. Given the following set rosters and the set diagram, find the sum of the numbers in the three shaded regions.

   $A = \{1, 2, 3, 4, 5, 6, 7\}$
   $B = \{1, 3, 5, 7, 9\}$
   $C = \{3, 4, 5, 6, 7\}$
   $D = \{-3, -1, 1, 3, 4\}$

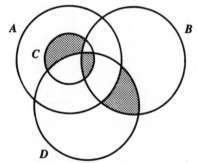

3. A farmer's wife drove to town to sell eggs. To the first customer she sold half her eggs and half of an egg. To the second customer she sold half of the remaining eggs and half of an egg. To the third customer she sold half of the remaining eggs and half of an egg. To the fourth customer she sold her last three eggs. If she did not break any eggs, how many did she sell?

4. A circle is inscribed in an equilateral triangle and a square is inscribed in the circle. What is the ratio of the perimeter of the triangle to the perimeter of the square?

5. Find the units digit in the final product when $(2543)^{54}$ is multiplied out.

6. What is the probability that a positive integer selected at random is relatively prime to 12? Express your answer as a common fraction in the lowest terms. (Relatively prime to 12 means the only common factor is 1.)

7. Mr. Whipple wants to blend two teas, regular tea having a wholesale cost of $0.90 per pound and spice tea at a wholesale cost of $1.20 per pound. He wants to make a profit of 50% over the wholesale cost and sell the tea for $1.65 per pound. At what ratio of regular tea to spice tea should he mix the two types?

8. Find two positive integers, each less than 8, such that the sum of their squares plus their product is a perfect square. Give your answer as the sum of the two squares plus the product.

9. A rowing team consists of eight men, one of whom can only row on the right and one of whom can only row on the left. In how many ways can the crew be arranged with four men on each side of the boat?

10. In the following figure $ACDF$ is a rectangle, and $\overline{BG} \perp \overline{CE}$. $AC = 4$, $AB = FE$, $CD = 2$, and $BC = \sqrt{2}$. Find BG.

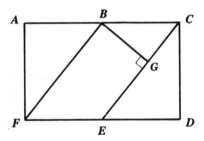

11. For what values of y will the equation $\sqrt{1 - x} + \sqrt{1 + x} = y$ have at least one real solution for x? Express your answer in the form $a \leq y \leq b$.

12. In the following diagram each square should contain a different digit from 1 through 9. If two (or three) squares are adjacent, then this will represent a two- (or three-) digit number. Determine the placement of the digits that will give a true statement. Ignoring the obvious symme-

tries, there are two possible answers. For your answer give the smaller of the three-digit numbers.

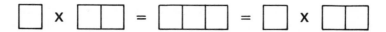

13. A circular table stands in a corner touching both walls. A point on the edge of the table is 8 inches from one wall and 9 inches from the other wall. Find the largest diameter of such a table.

14. Given that the ratio of the interior angles of two regular polygons is 3:2, how many such pairs are there?

15. On Armistice Day (November 11) 1928, General Foghorn T. Plunder claimed that he had lived as long in the twentieth century as he had lived in the nineteenth. What was his birthdate?
(To keep this problem from being too deceptive, it should be noted that January 1, 1901, was the first day of the twentieth century and 1900 was *not* a leap year.)

16. At what time between the hours of five and six will the hour hand and minute hand be at right angles? Give the first time of the two possible answers. Give an exact answer.

17. Determine the greatest value of k for which a root of the following polynomial function is -2:

$$P(x) = x^3 - (k + 3)^2 x^2 - 7x + k$$

18. Find the value of N:

$$N = \sqrt[3]{6 + \sqrt[3]{6 + \sqrt[3]{6 + \ldots}}}$$

19. In the figure, *AFCH* represents the vertices of a tetrahedron contained in a cube having edges of length 3. What is the volume of the tetrahedron?

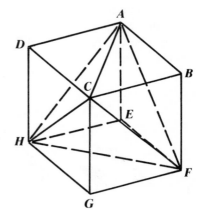

20. There are three prisoners: Alfred, Bixby, and Chester. The prisoner with the highest degree of guilt will be executed. Alfred sees the warden and asks for any information that he has. The warden says that from what he has reviewed, he is sure that Bixby will not be executed, but that he has not yet reviewed Alfred's case. Assuming that there are no ties in the degree of guilt, what is the probability that Alfred will get executed?

(Answers on page 111)

FRESHMAN–SOPHOMORE LEVEL

1. Little Alfred loves to build patterns with his blocks. In the diagram, one can see the beginning of Alfred's pattern. If he continues with the pattern until the base is fifteen cubes wide, how many cubes will he have used to make all fifteen "walls"?

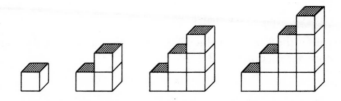

2. Alfred now tries to make a more complex pattern. For the front of the pattern he has steps leading up and down, but every time he increases the height by one additional row, he also increases the depth by building additional sets of steps. If the base of the pattern is to have fifteen cubes, then how many cubes will he have to use to make the largest "building"?

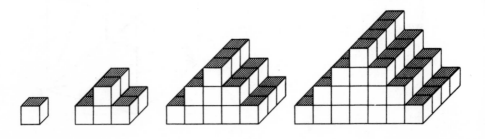

3. From 1970 to 1980, the population of Cook County increased by 3%. If the population of the suburbs increased by 8% and the population of the city of Chicago decreased by 1%, then what was the ratio of the suburban population to the city population in 1970?

4. Three years ago Tom was three times as old as his sister. In three years Tom will be twice as old as his sister. How old is Tom?

5. How many minutes before one o'clock will the minute hand and the hour hand be the same distance from the 12?

6. A solid cylinder has a conical core removed so that the base of the cylinder is the base of the cone, and the upper vertex of the cone is on the upper base of the cylinder. If the height of the cylinder is 10 and the volume of the part of the cylinder that is left after the cone is removed is 20π, what is the radius of the cylinder?

7. The Mix and Match Clothing Store sells all suit coats for a single price, all trousers for a fixed price, and all vests for a single price. A coat costs the same as two vests and a pair of trousers. Three coats and a pair of trousers cost $290. What is the price of a complete suit?

8. A contractor hires daily help in a rather unusual way. He is willing to pay a fixed amount for the total pay for all the workers, with this amount being divided evenly among the workers. If there were two fewer workers, each worker would receive $15 more per day. If there were four additional workers, each would receive $15 less each day. How much does the contractor pay per day?

9. In the following figure, $AB = 5$, $BC = 12$, and $AC = 13$. From a point P in the interior of the triangle, perpendiculars are constructed to each of the sides. If $PD = 1$ and $PE = 2$, find PF.

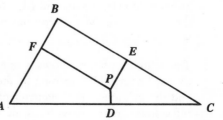

10. Find the positive integer solutions for $3x + 8y = 103$. Write your answer as ordered pairs of the form (x, y).

11. Shelley has six pieces of a chain, each consisting of four links. She would like to put these together to form a closed chain of twenty-four links. If the jeweler charges 25¢ to cut a link and 50¢ to reassemble the link, what is the least that it will cost to make the chain?

12. Albert goes into a store and says, "Lend me as much money as I already have and I will spend $20 in your store." The owner agrees to lend Al the money, and Al spends the $20. Al does the same thing at a second, third, and fourth store. After this he has no money left. How much does he owe the four store owners?

13. There are five weights such that A and B together weigh exactly the same as C and E together. B and E together weigh the same as A, C, and D together. A weighs more than B, and B weighs more than D. B and D together weigh more than B and C together. Which weight is the lightest?

14. An owner of a store buys a watch at a wholesale price of $15, then sells it for $30. A man who purchases the watch pays for the purchase with a $100 bill. When the store owner takes it to the bank, he is told that it is counterfeit. Therefore the bank will not accept the money. What is the total loss to the store owner?

15. A carpenter can cut through a board in 10 seconds. He can cut through a double thickness in 15 seconds and a triple thickness in 20 seconds. If he cannot cut through more than three boards at once, what is the minimum cutting time for dividing a long board into sixteen parts?

16. How many positive six-digit integers can be formed using
$$1, 2, 2, 3, 3, 4.$$

17. Four couples have been bowling for years, and between them they have won fifty-four trophies. Alice has won two, Betty has won three, Cora has won five, and Diane has won seven. Mr. White has won as many as his wife, Mr. Xnant and Mr. Youghart have won twice as many as their wives, and Mr. Zircon has won three times as many as his wife. Who is Mrs. Zircon?

18. The following figure shows a rain gauge (a truncated cone connected to a cylinder). The opening at the top is 60 centimeters in diameter, and the cylinder has a diameter of 20 centimeters. The opening along the side is a window with a scale to tell the depth of the rain. What is

the calibration scale? That is, for each centimeter of rain, how far would the water rise in the cylinder?

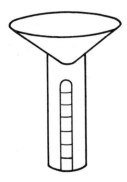

19. One root of the quadratic $2x^2 - 5x + q = 0$ is 4. What is the other root?

20. A plane travels from here to Los Angeles and back. The plane flies 600 miles per hour in still air, but it has a 50-mile-per-hour head wind going and a 60-mile-per-hour tail wind returning. What is the average speed over both trips?

(Answers on page 111)

JUNIOR–SENIOR LEVEL

1. Two gears are turning together. The smaller gear has fifteen teeth and the larger gear has twenty-one teeth. A point on the first gear is matched with a point on the second gear. After how many turns of the smaller gear will the points again match?

2. Tom has designed an eight-hour clock. The minute hand still takes sixty minutes to complete one revolution, but for the complete revolution of the minute hand, the hour hand goes one-eighth of a revolution (rather than one-twelfth). At how many minutes after one o'clock will the minute hand and the hour hand coincide?

3. From 1970 to 1980, the population of Cook County increased by 3%. If the population of the suburbs increased by 8% and the population of the city of Chicago decreased by 1%, then what was the ratio of the suburban population to the city population in 1980? (Hint: Let the populations be some arbitrary number.)

4. Solve for all real values of x:
$$x(x + 2)(x + 1)(x - 1) = 120$$

5. If $0 < x, y, z$, solve for (x, y, z):
$$xy = 30 \qquad xz = 10 \qquad yz = 3$$

6. Twelve times a number is greater than 7 by the same amount that four times the number is greater than 23. What is the number?

7. Solve for x:
$$x^2 = 21 + 4\sqrt{5}$$

8. A right pyramid has a height of 9 meters. Two planes cut through the pyramid parallel to the base such that they cut the pyramid into three pieces of equal volume. What is the distance between the planes?

9. Find the value of N for the following infinite continued fraction:

$$N = 1 + \cfrac{1}{2 + \cfrac{1}{3 + \cfrac{1}{1 + \cfrac{1}{2 + \cfrac{1}{3 + \cfrac{1}{1 + 1}}}}}}$$

10. Solve for (x, y):
$$x^2 - 5xy + 6y^2 = 72$$
$$6x^2 - 5xy + y^2 = 72$$

11. In how many ways can you have 25¢ in standard United States money (pennies, nickels, dimes, and quarters)?

12. If $(a + \frac{1}{a})^2 = 3$, find $a^3 + \frac{1}{a^3}$.

13. Find all the positive integral solutions for which $x \leq y$. Write your answer as ordered pairs (x, y).
$$\frac{1}{x} + \frac{1}{y} = \frac{1}{4}$$

14. A square grid is made of very, very thin wire. The squares formed are 5 cm on an edge. A coin 2 cm in diameter is dropped on the grid. What is the probability that the coin will pass through without touching a wire?

15. In how many ways can five distinct objects be distributed among three individuals? (Note: Giving all five objects to one person is considered to be one way of "distributing" the objects.)

16. Find the area of a triangle whose sides measure 6, $\sqrt{2}$, and $\sqrt{50}$.

17. If sin A = 0.96, sin B = 0.44, and sin C = 0.62, then find 4 cos $\frac{A}{2}$ cos $\frac{B}{2}$ cos $\frac{C}{2}$. Your final answer should be expressed as a decimal real number.

18. The object of a game is to roll and eventually match three dice. If all three are the same, it is a winning roll. If two are the same, then the die that is different is rolled again. If it matches, this is a win. If all three are different, then all three are rolled again, and if they match on the second roll, this is also a win. What is the probability of winning in this game?

19. The perimeter of a right triangle is 36. The sum of the squares of the three sides is 450. Find the length of the shortest leg.

20. An algebraic quotient for which the division of the numerator by the denominator produces an infinite series is called a compact quotient; for example, $1/(1 - x) = 1 + x + x^2 + x^3 + \ldots$. Find a compact quotient that produces the series $1 + 2x + 3x^2 + 4x^3 + \ldots$.

(Answers on page 112)

Relay Team Competition

As with the eight-person teams, there is opportunity for flexibility in establishing relay teams. Teams may be separate for each grade level, consist of students from each grade level, or consist of students from one of two levels—freshman-sophomore or junior-senior. Each school may be allowed to enter one, two, or three teams.

A team consists of four persons. For example, if there are two levels, the freshman-sophomore team should have two freshmen and two sophomores, and the junior-senior team should have two juniors and two seniors.

During the competition a team sits in a row with the freshmen sitting in front of the sophomores or the juniors in front of the seniors. At the freshman-sophomore level, geometry questions should be restricted to the sophomores, and at the junior-senior level trigonometry questions should be restricted to the seniors.

There are usually three rounds of competition, but you can have as many rounds as you wish. It is merely a matter of time. Each round takes about ten minutes to complete.

Each student should have at least three official answer sheets prior to each round. An official answer sheet can be simply a small sheet with a space for the school name and four blanks like the following:

Freshman-Sophomore Relay

School: _____

1. _____

2. _____

3. _____

4. _____

Score: _____

The proctors distribute, face down, the appropriate question to each person. (Having designs on the back of the problem sheets keeps the contestants from reading the problem through the sheets.) At the signal to begin, each person turns over his or her problem. The first member of the team solves a problem, puts the answer on the official answer sheet, and passes it to the second member of the relay. This answer should be used in the statement of the second person's problem. Problems 2, 3, and 4 usually begin with "Let k = ANS," and the letter k is used in the context of the problem.

A team will not submit an answer sheet until all four contestants have answered their questions. When a team has all four answers, the contestant in the fourth seat awaits the command to stand and present the four answers. Each contestant has three answer sheets to allow an opportunity to change an answer if he or she finds an error after an answer was passed back. If an answer is changed, it affects all the subsequent questions, and a new answer sheet has to be passed. Eventually the person in the fourth seat must submit the most current answer. Note: No information other than answers should be passed during the competition.

For each round the first command to stand will be given at the end of three minutes, the second at the end of five minutes, and the third at the end of seven minutes.

Experience has shown that getting all four answers correct, even in seven minutes, is no minor task. If the four problems are reasonably easy, for example, having a difficulty index of .80, then the probability of getting all four problems correct is .36, and this decreases dramatically if the problems are slightly more difficult. Therefore, it is recommended that points be given if some of the problems are answered correctly.

Scoring on official answer sheets should be as follows:

Question 1 correct:	1 point
Questions 1 and 2 correct:	2 points
Questions 1, 2, and 3 correct:	3 points
All four questions correct at first command to stand:	10 points
All four questions correct at second command to stand:	7 points
All four questions correct at third command to stand:	5 points

It is assumed that the correct answer in one position is necessary for the subsequent positions. Therefore if a team answers question 1 correctly, misses question 2, but somehow manages to answer questions 3 and 4 correctly, they would receive only 1 point.

The following examples consist of three rounds of questions for teams with one student at each grade level, six rounds of questions for freshman-sophomore teams, and six rounds of questions for junior-senior teams. Notice how the answer to each question is needed for subsequent questions.

FOUR GRADE LEVELS

Contestants solve problems for their own grade level.

Round 1

1. (Freshman) If $a{:}b{:}c = 1{:}3{:}5$, then find: $\dfrac{a + 3b + 5c}{a}$

2. (Sophomore) Let $k = $ ANS $- 5$. *All* of the triangles in the figure and the central hexagon are equilateral. If $AC = k$, then find the area of the entire star.

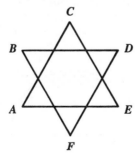

3. (Junior) The answer you receive should be of the form $a\sqrt{b}$. Let $k = 2b$. Solve the following for x:

$$\frac{0.3x - k}{0.5x - 2.4} - \frac{3 + 1.2x}{2x - 0.1k} = 0$$

4. (Senior) Let $k = $ ANS. Find the area of the region determined by the following inequalities:

$$y \geq x, y \geq 1, \text{ and } x^2 + y^2 - 2x - 2y + 2 \leq k^2$$

Round 2

1. (Freshman) There are two numbers differing by 10 whose sum is equal to twice their difference. Find the larger of the two numbers.

2. (Sophomore) Let $k = $ ANS. M is the midpoint of minor arc AC and B is the midpoint of the major arc AC. If $m\angle MBC = k°$, then find $m\angle ACB$. (Do not write the degree symbol in passing your answer.)

3. (Junior) Let $k = $ ANS $+ 5$. The sum of two numbers is k. The sum of the squares of the two numbers is 16 more than twice the product. What is the larger number?

4. (Senior) Let $k = $ ANS. Find the numerical coefficient of the x^{25} term in the expansion of

$$\left[x^2 - \frac{k}{7x}\right]^{14}.$$

Round 3

1. (Freshman) Given that the cubic equation $x^3 + ax^2 + bx + c = 0$ has solutions $-2, 1, 3$, then what is the value of a?

2. (Sophomore) Let $k = |\text{ANS}|$. In the following figure, $ABCD$ is a rectangle inscribed in a semicircle. If $AB = 2BC$ and the radius of the circle is k, find the area of the segment of the circle determined by $\overline{AB}$.

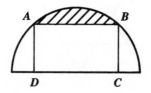

3. (Junior) Let $k = $ ANS $- \pi$. Find the value of

$$\frac{4 + k\sqrt{3}}{4 + k\sqrt{3}} + \frac{4 - k\sqrt{3}}{4 + k\sqrt{3}}.$$

4. (Senior) Let $k = $ ANS. Solve the following for x:
$$\sqrt{kx + 3} + \sqrt{kx - 3} = \sqrt{6}$$

(Answers on page 112)

FRESHMAN–SOPHOMORE LEVEL

Freshmen solve problems 1 and 2.
Sophomores solve problems 3 and 4.

Round 1

1. If 13 hens lay 25 eggs in 3 days, how many hens does a farmer need to get 400 eggs in 6 days?

2. Let $k = $ ANS.
   If $X^2 + Y^2 = k$
   and $2XY = 40$, what is $|X + Y|$?

3. Let $C = $ ANS.
   A 30-60-90 triangle has hypotenuse C. What is its area?

4. Let $X = $ ANS.
   Simplify:
   $$\frac{X + \sqrt{3^5}}{X - \sqrt{3^5}}$$

Round 2

1. What is the x-intercept of $y = 9x - 12$?

2. Let $R = $ ANS.
   Find X if $\dfrac{10}{X} + \dfrac{20}{X} + \dfrac{30}{X} + \dfrac{40}{X} = R$.

3. Let $G = $ ANS.
   Find the positive mean proportional between 3 and G.
   (That means $3:X$ and $X:G$ are in the same ratio.)

4. Let P = ANS.

If the perimeter of this isosceles $60°$ trapezoid is P and the shorter base is the same length as the sides, what is the length of the longer base?

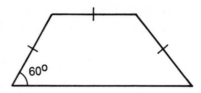

Round 3

1. The sum of six times a certain number and five times the number is 143. What is the number?

2. Let d = ANS.

The difference of the squares of two consecutive positive integers is d. What is the larger of the two integers?

3. Let F = ANS.

X is the smallest number such that $X > 120$ and F is a factor of X. Find $\dfrac{X}{F}$.

4. Let C = ANS.

The area of a rectangle is C square inches. The perimeter of the same rectangle is C inches. What is the length in inches of the shorter side of the rectangle?

Round 4

1. Determine the value of the following expression:
(The symbol "$\times$" means to multiply.)
$$2 \times (3 + 6) - 3 \times (4 - 2)$$

2. Let n = ANS. The difference between the ages of Mary and Jack is n years. In two years Mary will be four times as old as Jack. How old is Mary?

3. Let n = ANS. A circle of diameter n has 6 points placed evenly on the circle. (That is, the distance between any two consecutive points, including the first and sixth, is equal.) What is the distance between the first and fifth points?

4. Let n = ANS. In the following figure, $ABCD$ and $EFGH$ are congruent, coplanar rectangles and $m\angle BPE = 90°$. If $AD = n$, $AB = 3\sqrt{3}$ and $EP = \sqrt{3}$, find the area of the shaded region.

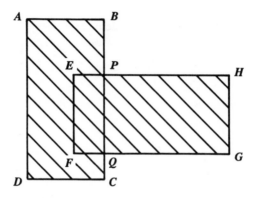

Round 5

1. Pick any integer between 3 and 7. Square the number, then add it to six times the number. Pass the result to the person in the second seat.

2. Let n = ANS. Determine two integers p and q such that $p = 3n + 5$ and $q = 6n^2 + n - 9$. Divide q by p. Pass the remainder back to the third student.

3. Let k = ANS. In the following figure, $AC = k + 4$, $AD = k - 1$, and $DE = \dfrac{k}{2}$. Find BC.

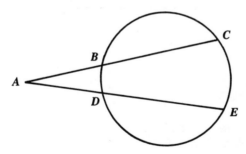

4. Let s = ANS. A triangle has a base of $5s$. A trapezoid has the same altitude and the same area as the triangle. If one base of the trapezoid is seven times as long as the other base, find the length of the median of the trapezoid.

Round 6

1. Find the arithmetic mean (average) of the positive even integers from 2 through 100.

2. Let n = ANS. Let n be the arithmetic mean (average) of 50 consecutive even integers. What is the largest of these 50 integers?

3. Let n = ANS. In the following figure, $AB = AD$ and $AC = BC$. If $m\angle ADC = n$, find the measure of $\angle C$.

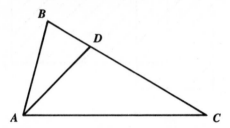

4. Let n = ANS. In the following figure, $ACDF$ is a rectangle and $BCDE$ is a square. If $AB = n$, and $m\angle AFC = 60°$, find AF.

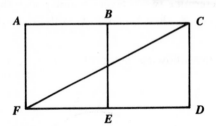

(Answers on page 112)

JUNIOR–SENIOR LEVEL

Juniors solve problems 1 and 2.
Seniors solve problems 3 and 4.

Round 1

1. Find the y-intercept of the line parallel to $7x + 6y = 4$, which contains the point $(6, -1)$. (The y-intercept is the y-coordinate of the point where the line crosses the y-axis.)

2. Let k = ANS. A stick is first cut in half, then one piece is cut in a ratio of 2:1. Of the three pieces, the total length of the shortest and the longest pieces is k. How many units long was the original stick?

3. Let $k = \dfrac{\text{ANS}}{3}$. Then k is the geometric mean between -2 and what number?

4. Let $k = \dfrac{-2}{3}$ ANS. Find the value of $(-\sqrt{3} + i)^k$. Give your answer in $a + bi$ form.

Round 2

1. Assume that you have 24 kilograms of water. One liter of water has a mass of one kilogram and one liter is equivalent to 1000 cubic centimeters. A rectangular container having a width and height both equal to ⅓ of the length is just large enough to hold this water. Find the length in centimeters.

2. Let k = ANS. The perimeter of an equilateral triangle is k, and its vertices are $P(0, 0)$, $Q(A, 0)$, and $R(B, C)$, where A, B, and C are all positive. Find the value of b if $\overleftrightarrow{QR}$ crosses the y-axis at $S(0, b)$.

3. The answer you receive will be of the form $a\sqrt{b}$. Let $k = a - 4b$. Find the seventh term of the geometric series whose first two terms are

$$\sqrt[3]{k}, \ \sqrt{k}, \ldots$$

4. Let ANS be the number of cubic centimeters in the volume of a right circular cylinder. The number of square centimeters in the area of the top is equal to the height in centimeters. Find the height in centimeters.

Round 3

1. Find the length of the side of a square whose area is equal to a rhombus having diagonals of length 12 and 6.

2. Let $k = \dfrac{\text{ANS}}{3}$. Solve for the largest positive root of the cubic equation

$$2x^3 + x^2 - 4kx = 2k.$$

3. Let k = ANS. Let $f(x) = k$, and let $g(x) = \sqrt{x}$. Find:

$$f[g(x)] + g[f(x)]$$

4. The answer you receive should be in the form $m + \sqrt{n}$. Find the maximum y-coordinate of the intersection points of the circle and line given by the equations:
$$x^2 + (y - m)^2 = n^2 + 1$$
$$y = nx + m$$

Round 4

1. Simplify the following: $\dfrac{m^2 - 1}{2m^2 - m - 3} \cdot \dfrac{4m - 6}{m - 1}$

2. Let $n =$ ANS. Then n is the geometric mean between 4 and k, and k is the arithmetic mean between 4 and m. Find m.

3. Let $k =$ ANS. Evaluate the following determinant:
$$\begin{vmatrix} 1 & 1 & 1 \\ -1 & -1 & k \\ 2 & k & 2 \end{vmatrix}$$

4. Let $n =$ ANS. Assume $\log_{10} 3 = 0.48$ and $\log_{10} 5 = 0.70$.
Find $\log_{10} \left(\dfrac{1}{45}\right)^n$.

Round 5

1. Find the sum of the roots of the lowest common multiple of the following polynomials:
$$x^2 - 6x + 8 \qquad x^2 - 7x + 10 \qquad x^2 - 4x + 4$$

2. Let $k =$ ANS $- 1$. If k is added to a number, the result is the same as 45 times the reciprocal of the number. What is the largest such number?

3. Let $k =$ ANS. A bag contains five red marbles and five white marbles. If k are selected from the bag, without replacement, then what is the probability that at least one of the marbles is red?

4. The answer you receive should be of the form $\dfrac{p}{q}$. Let $\cos A = \dfrac{p - 7}{q - 7}$ and $\tan B = \dfrac{p - 8}{q - 8}$. Find $\sin (A + B)$. Assume that A and B are angles in quadrant I.

Round 6

1. What is the area of a circle with center at $(-2, 3)$ and containing the point $(4, 11)$?

2. Let n = ANS. A circular garden is surrounded by a uniform circular walk having a radial width (the width of the walk along a radius) of 4. If n is the area of the uniform circular walk, what is the diameter of the garden?

3. Let n = ANS. Both of the roots of the quadratic equation $x^2 + 4x + k = 0$ are integer factors of n and have a product of n. What is the value of k?

4. Let n = ANS. If $\cos \theta = \dfrac{n}{35}$, $0 < \theta < \pi$, find $\sin (\pi - 2\theta)$.

(Answers on page 113)

Two-Person Team Competition

As with the relays, there are several ways to organize a two-person team. For purposes of discussion, it will be assumed that a two-person team will represent either a freshman-sophomore level, or a junior-senior level. However, it is possible to have two-person teams at each grade level. If you want even more participation, there can be two or three divisions.

For this competition, a problem is presented using an overhead projector and the teams try to solve the problem as quickly as possible. Contestants on a team may work together. As soon as they have an answer they hold an answer sheet high in the air, and the answer sheet is then collected by a proctor. Each round has a three-minute time limit. For each round, the first team to have the correct answer receives 7 points, the second team receives 5 points, and all other teams that have a correct answer within the three-minute time limit receive 3 points each.

Experience has shown that there should be no more than twelve two-person teams in any room. Therefore, if there are many teams competing, it is advisable to break them into divisions. At times it is very difficult to determine the order in which teams submit their answers. It is best simply to collect the answers in sequence and not to score the answers until later. For example, there may be only a one-second difference between the time the third and fourth answers are submitted. If the first two answers submitted are determined to be incorrect, this one-second difference becomes the difference between first and second place in the round.

This format has also been used for entertainment during an awards presentation. For example, suppose there were three rooms used for the competition. While the scores are being compiled for the awards, the top teams from each room can have a demonstration competition. The first team to win three rounds is then the overall winner. The demonstration often proves to be successful because all the students have an opportunity to try the problems. It is much like watching a game show on television.

Although the example problems that follow were used in competitions, it should be noted that some care must be taken in preparing questions for a contest. The problems will be shown on a screen, which has to be seen by all the contestants in the room. The screen should be situated so that there is no distortion in the problem, and the problem must be written in large print. Putting each problem on a separate transparency and using an elementary typewriter have proved successful. Recently, high-quality computer word processors and printers have made the task much easier.

FRESHMAN-SOPHOMORE LEVEL

1. Which is the largest?
$$2^{27}, \qquad 3^{18}, \qquad 5^{12}$$

2. Two wheels, of radii 2.5 meters and 3.0 meters, are rotating in contact with each other. At the beginning, an arrow is drawn on each wheel at the point of contact. How many revolutions must the smaller wheel make before the arrows are next in contact?

3. Let $U = \{5, 12, 15, 17\}$. If $A = \{5, 12\}$ and $A \cap B = \{5\}$ and $A \cup B = U$, find the sum of the numbers that are elements of the set B.

4. What three-digit number is equal to eleven times the sum of its digits?

5. Andy is as old as Ben and Charlie together. Two years from now, Andy will be twice as old as Charlie. Last year Ben was twice as old as Charlie. What is the sum of their ages?

6. If $\dfrac{x^n(x^{n+1}y^{2n-1})^3}{(x^{2n-1} \, y^{3n-2})^2} = x^r y^s$, find r.

7. The edge of a pond is a perfect circle. A fish starts at one edge, swims due north 60 meters, at which point he touches the edge of the pond again. He then swims 80 meters due east and again touches the edge of the pond. The radius of the pond is therefore how many meters?

8. Find the surface area of a regular tetrahedron with edge $\sqrt{2}$.

9. A pipe that takes thirty minutes to fill a tank is shut off after the pipe has been running for ten minutes. A second pipe is then opened, and it finishes filling the tank in fifteen minutes. How many minutes would it take the second pipe to fill the tank by itself (that is, from empty to full)?

10. For $x > 0$, let $f(x) = \dfrac{x}{1 + x}$. Find the least possible value of $B - A$, where $A < f(x) < B$ and A and B are integers.

11. Two rectangles are similar and the diagonal of the smaller rectangle is equal to the length of the larger rectangle. If the ratio of the length to the width of each of the rectangles is 3:2, what is the ratio of the area of the larger rectangle to the area of the smaller rectangle?

12. Find the smallest value of k for which this equation in x has real roots:
$$x^2 - kx - 4 = kx^2 + 2x$$

13. When the following complex fraction is reduced to a simple fraction, what is the numerator if the denominator is $(2x + y)$?

$$\frac{1 - \left(\dfrac{1}{1 - y/x}\right)}{1 - \left(\dfrac{3}{1 - y/x}\right)}$$

14. The sum of the areas of two similar polygons is 68. Their corresponding sides are in the ratio of 3:5. Find the area of the smaller polygon.

15. If one solution to the following polynomial function is 2, find the larger of the other two solutions.
$$x^3 - 3x^2 - 4x + 12 = 0$$

16. Add the following series:
$$\frac{1}{2} + 1 + 1\frac{1}{2} + 2 + 2\frac{1}{2} + \ldots + 19\frac{1}{2} + 20$$

17. If you increase an item by 25%, by what percent must you reduce it to bring the price back to the original price?

18. A baseball team wins forty of its first fifty games and has forty remaining games to play. How many of the remaining games must the team win in order to win 70% of its entire schedule?

19. If a boat travels 20 miles per hour in still water and takes three hours longer to go 90 miles against the current than it does to go 75 miles with the current, what is the speed of the current of the river?

20. Determine the maximum value of the function f defined by
$$f(x) = -2x^2 - 7x - x.$$

(Answers on page 113)

JUNIOR-SENIOR LEVEL

1. Let L be the line in which the planes $2x + y - z = 13$ and $x - 2y + z = -4$ intersect. If the point $(A, 3, B)$ lies on L, how much is $(A - B)$?

2. The minute hand on a clock is 12 centimeters in length and the hour hand is 6 centimeters. What is the ratio of the linear speed of the tip of the minute hand to the linear speed of the tip of the hour hand?

3. Find t, if $x + 1$ is a factor of
$$3x^3 - 2x^2 + tx - 4.$$

4. What is the value of $\cos\left(\arctan\left[\dfrac{-12}{5}\right]\right)$ in simplest form?

5. A train traveling at 30 meters/second passes a post in 20 seconds and passes through a 300-meter tunnel in 30 seconds. How many meters long is the train?

6. Find: $\log_{1/64}(16\sqrt[3]{16})$

7. Simplify completely:
$$\cos(2x) + 2\sin^2 x$$

8. The medians to the legs of a right triangle are 5 and $2\sqrt{5}$. What is the length of the hypotenuse?

9. Solve for x in radians, with $0 < x < \pi/2$:
$$4\sin(\pi/2 + 2x) - 2\sin x = 1$$

10. What is the value of $\log_8(2\log_{10}100)^2$?

11. Find the sum of this arithmetic series:
$$\sum_{k=1}^{18}\left(2\dfrac{2k - 1}{3}\right)$$

12. What is the value of $e^{3\ln 2 + \ln 3}$ in simplest form?

13. The ratio of the volumes of two cubes is 9:1. Find the ratio of the urface areas. State answer as the ratio of the larger to the smaller.

14. The bases of an isosceles trapezoid are 10 and 16. The nonparallel sides are each 6. Find the area of this trapezoid.

15. Find the period of the function:
$$F(t) = 2\sin(5t) - 3\cos(5t)$$

16. Find $a^2 + b^2 + c^2$ where a, b, c are real numbers and for all x:
$$(x - a)(x - b)(x - c) = x^3 - 2x^2 - 9x + 18$$

17. Find the smallest number that is *not* a solution to this inequality:
$$\left|\frac{x + 4}{x - 3}\right| < 2$$

18. If one die is fair and the second die is "loaded" so that a "6" is twice as likely as any other number, what is the probability of rolling a "7" with the sum of the dice?

19. A point P is located in the interior of an equilateral triangle whose side lengths are $12\sqrt{3}$. What is the sum of the distances from P to the three sides?

20. If x is 2 greater than y, and y is 13 more than 8 times 13, find the average of the values x, y, and $(3x - 2y + 1)$.

21. $\left(27^{2/3} + 64^{2/3}\right)^{3/2} - 10^2 = ?$

22. A, B, C, D are four distinct numbers.
$$D + A = D$$
$$A \cdot D = A$$
$$B + C = A$$
$$B(A + D) = D$$
$$B - C = D$$
$$D \text{ equals } \underline{\quad ? \quad}$$

23. How many integers K, $1 \le K \le 206$, are multiples of 4 or 6?

24. How many positive integers (x) less than or equal to 100 have the property that x is a multiple of 3 or x is a multiple of 5?

25. Find the coefficient of x^5 in
$$(x^5 - 2x^4 + 3x^2 - x + 3)(x^6 + 3x^5 - 4x^3 + 2x^2 + 3x - 4).$$

26. If $f(x) = \sqrt{x^2 + 8x + 16} - \sqrt{x^2 - 8x + 16}$, $f(15.73) = ?$

27. Write the sum of the absolute values of the coefficients of the quotient of $\dfrac{x^5 - 1}{x - 1}$.

28. A line with slope 3 goes through the point $(8, 12)$. If the point $(C, -3)$ is on the line, find C.

29. Express the recurring decimal $2.345345345 \ldots$ as a reduced rational number.

30. Find all solutions (x, y) of the following system of equations. From the

set of solutions, what is the maximum value of y that corresponds to the maximum value of x among the set of all solutions?

$$x + y = 2$$
$$x^2 + y^2 = 20$$

(Answers on page 113–14)

Calculator Competition

As calculators (and computers) are more commonly accepted as tools for *doing* mathematics rather than *bypassing* mathematics, skill in using calculators has become a potential topic for competition. The wide array of calculators available makes problem writing a difficult task unless there are some restrictions placed on the quality of the calculator. For example, if there is a calculator that will solve a system of up to four linear equations simply by entering the coefficients and constants, a student having this calculator has an advantage over a student who must solve the system using Cramer's rule. There is another reason to restrict the quality of the calculator. The cost of a sophisticated calculator might be prohibitive to some students. Limiting the quality of the calculator also limits its cost.

There are several possible ways to group students for a calculator competition: individuals at each grade level, individuals from two divisions (freshman-sophomore and junior-senior), individuals from any grade level in the school, or teams from any of these categories. The examples which follow will be from several categories.

Time limit is a function of the format. For the individual formats, each student can be expected to do ten problems in thirty minutes. A team of five students could be expected to do twenty problems in twenty minutes, but several of these problems could be quite complex. There is certainly considerable flexibility in organizing this competition, and participating schools can experiment with various formats.

Two major concerns are the form in which an answer should be written and the scoring, particularly if an answer is partially correct but is incorrect at the third, fourth, or fifth significant digit. Here is a suggested format for a contest:

All answers must be written legibly in the correct place on an answer sheet. Each answer must be written in scientific notation ($a \times 10^n$, where n is an integer and $1.0 \le a < 10.0$) and rounded to four significant digits. The rule for rounding is that if the fifth digit is greater than or equal to 5, then the fourth digit is rounded to the next integer; and if the fifth digit is less than 5, it will be truncated.

In scoring, there are two possible techniques: full credit or systematic partial credit. One possible technique for giving partial credit would be to assign points depending on the number of significant digits that are correct: if correct to one significant digit, 1 point; two significant digits, 2 points; three significant digits, 3 points; and four significant digits, 5 *points*. Points can be deducted if the answer is not written in correct scientific form, does not have four significant digits, or omits some aspect of the answer. For example, if the correct answer is 1.320×10^2, an answer of 1.32×10^2 would be worth 3 points because it is correct to only three significant digits, and a point might be deducted simply because it does not have four significant digits listed. Another example is the omission of units when appropriate. If the correct answer is 4.375 km, an answer of 4.375 would have a point deducted.

It might be worth noting that although a computer competition has real potential, at this time it is not reasonable. The variation in computers is much greater than the variation among calculators, and in a competition it is reasonable to restrict the contestants to identical equipment. If this problem can be overcome and a computer competition is held, questions should be similar to those used in an oral competition or a two-person power team. That is, the problems should be mathematically sophisticated, the students should prepare for the topic, and the solutions should be evaluated for quality as well as correctness. At this time, however, it is unlikely that a direct competition of computer skills could be done effectively.

In the examples that follow it is assumed the students have a calculator that cannot be programmed and that has a maximum of two memory levels. The calculators are assumed to have typical scientific notation—roots, inverses, and trigonmetric, logarithmic, and exponential functions. The answer will be given in standard decimal notation to four significant digits.

Leave all answers rounded to four significant digits, unless otherwise stated in the problem. All angles are in *radians* unless otherwise stated.

1. Solve for x: $215 = e^{\left(x + \frac{4.723}{2}\right)}$

2. If all angles are in radians, find
$$\sin 2.65 + \cfrac{1}{\sec 3.81 - \cfrac{1}{\tan 4.78 + \cfrac{1}{\csc 5.24}}}$$

3. Find the eighth term of a geometric sequence whose sixth term is 4 and whose tenth term is 5.

4. If $\log_{10} t + \log_{10} 3t - \log_{10} (t + 2) = 0.5$, then $t = $ _____.

5. Evaluate $\dfrac{\sqrt{1.2} + 4.3^{1.78}}{e^{1.57}}$.

6. To what number does the iterative procedure $x_0 = 1$, $x_{n+1} = \dfrac{1}{2}(x_n + \dfrac{2}{x_n})$ converge?

7. Evaluate: $x^{x^{x^x}}$ for $x = \sqrt{2}$

8. The following is an infinite geometric series for which the terms get increasingly smaller, eventually having little effect on the sum. Find the sum to three decimal places.

$$\frac{11}{23} + \frac{11}{23}\left(\frac{2}{7}\right) + \frac{11}{23}\left(\frac{2}{7}\right)^2 + \ldots$$

9. Solve for the positive root: $x\sqrt{\ln (\pi + e)} + ex^2 + \pi e = \pi x^2$

10. Two sides of a triangle are 1.342 and 2.947. The included angle is $\dfrac{7\pi}{12}$ radians. Find the angle opposite the side of length 2.947. Your answer is to be in radians.

11. If $f(x) = x \ln (x^2 + e^x)$ and $g(x) = \log_{10} (x^4 + 3.412)$, find $f(g(1.071))$.

12. Solve for x in radians: $|\cos(x + 1.293)| \div \cos x = 1.391$

13. Define $a \otimes b = a \cdot b + b$. Evaluate $x \otimes (y \otimes z)$ where

$$x = 1.7$$
$$y = 2.3$$
$$z = -1.8$$

14. Define $f(x) = x^2 + e^{-x}$. Find $f(f(f(1.2)))$.

15. Solve for x:

$$\frac{3.71x}{\pi(1.44)^2} = \frac{2.73x}{\pi} - 2\sqrt{82.3^2 + 9.75^2}$$

16. If a polynomial function is positive at one point and negative at another, then at some point between, the function must be 0 (a root). For the following polynomial function, find a root that is between 3 and 4. Give the correct answer to three decimal places.

$$2x^4 - 3x^3 - 21x^2 + 18x + 16$$

17. Evaluate to seven decimal places:

$$3 + \cfrac{1}{7 + \cfrac{1}{16 + \cfrac{1}{1 + \cfrac{1}{293}}}}$$

18. Find, to four decimal places, the area of a regular pentagon with side of $\sqrt{17}$.

19. Using the approximation $\ln(1 + x) = x - \dfrac{x^2}{2} + \dfrac{x^3}{3} - \dfrac{x^4}{4}$, calculate $\ln 0.718$ to five decimal places.

20. Find the acute angle θ, in radians such that $\csc \theta = \tan \theta$.

21. The volume of a sphere is 1000 cm^3 and is equal to the sum of the volumes of a cube and right circular cone. If the edge of the cube and the height of the cone are each equal to the radius of the sphere, what is the radius of the cone?

22. Find, if any exist, the smallest positive root of $5x^3 - 4x^2 - x - 1 = 0$.

23. For x *real* and k a nonnegative integer, let
$$\binom{x}{k} = \frac{x(x - 1)(x - 2) \ldots (x - k + 1)}{k!}.$$

Find $\binom{3.240}{8}$.

24. Evaluate $0.7489 \cos \theta - 2.495 \sin \theta$ given that $\cot 2\theta = -1.748$ and $0° \le \theta \le 90°$.

25. Find θ in radians if $\sin \theta + 0.1472\sqrt{\sin \theta} = 0.024\,70$ and $0 \le \theta \le 2\pi$.

26. If $a = 7.631$, $b = -8.254$, $c = -2.613$, $x = \sqrt[3]{a^2 - b^2}$ and $y = \sqrt[4]{b^2 + c^2}$, find the value of $x^2 - y^3$.

27. Find the exact value of
$$\begin{vmatrix} 4 & 1 & 3 \\ 5 & -1 & 4 \\ 6 & 4 & -3 \end{vmatrix} - \begin{vmatrix} -2 & 3 & 7 \\ 3 & -8 & 4 \\ 1 & 5 & 6 \end{vmatrix} \cdot \begin{vmatrix} 2 & 4 & 5 \\ 3 & 7 & -3 \\ -1 & 2 & -4 \end{vmatrix}$$

28. Write $.031\ 63_{10}$ in base 5 (to four significant base 5 digits).

29. Find the area of a triangle with sides of length 31, 13, 39.

30. The volume, V, of a cone is found using the formula $V = (1/3)\pi r^2 h$, where r is the radius of the base of the cone and h is the height. A cone, set on its base, is filled to half its height with water. If the diameter of the base is 18.4 m, the height of the cone is 12.7 m, and water weighs 9810 newtons/m^3, then what is the weight of the water in the cone?

(Answers on page 114)

Estimation Competition

To avoid the problem of determining the quality of the calculators allowed in a competition, one approach is to have a competition for which the contestants must have similar mathematical insight as needed in a calculator competition but for which they do not require the hardware. Estimation problems can be the basis for this type of competition.

In this competition, a contestant estimates the final answer and is given credit according to the following criteria. Answers are expressed in scientific notation. An answer within two units of the first significant digit *and* the correct order of magnitude is worth 3 points; if it is also within two units of the second significant digit, it is worth 4 points, and if it is within three units of the third significant digit, the answer is worth 5 points.

For example, if the correct answer is 3.45×10^6, an answer of 2.95×10^6 is worth 3 points, an answer of 3.29×10^6 is worth 4 points, and an answer of 3.48×10^6 is worth 5 points.

The examples that follow are divided into two levels: freshman–sophomore and junior–senior.

FRESHMAN–SOPHOMORE LEVEL

Directions: Give your best estimate for the answer to each problem. No calculators or tables are permitted. Bring only a pencil; scratch paper will be furnished. Units of measure are not required on answers. All answers must be expressed to three significant digits using scientific notation, $a \times 10^n$, where $1.0 \le a < 10$ and n is an integer. If the answer is $a \times 10^0$, the 10^0 factor may be omitted.

1. Simplify: $93.875^2 - 6.125^2$

2. Simplify:
$$1 + \cfrac{3.3}{\cfrac{7.8}{2.4 - 9.8 - 6.1}}$$

3. Evaluate: $\sqrt{56.8}$

4. Evaluate: $\sqrt{383\ 000\ 000}$

5. Solve for the largest solution:
$$3x^2 - 2x - 3 = 0$$

6. Assume the temperature of your house is normally 72°. The cost of heating reduces about 2.3% for each degree the temperature is lowered. What is the percentage saved if the temperature is reduced to 67°?

7. Find the area of a triangle with sides of length 6, 8, and 12.

8. If 2.54 centimeters = 1 inch, how many cubic inches are there in a block 25 centimeters by 47 centimeters by 40 centimeters?

9. If $xy^2 = 1$, and x is increased by 17.8% and y is decreased by 17.8%, then what is the value of xy^2?

10. Find the radius of a circle whose circumference is 12.

11. Find the radius of a circle whose area is 12.

12. If inflation is 8% per year, then how much will a $4.25 steak cost in ten years?

13. Evaluate: $\sqrt{\dfrac{(3.95 \times 10^8)(7.55 \times 10^2)^3}{35.7 \times 10^{-3}}}$

14. In an electrical circuit the relationship among voltage (E), current (I), and resistance (R) is given by the formula $E = IR$. If four resistances (r_i) are connected in parallel, the total resistance, R, is found by combining the four individual resistances in the following way:
$$\frac{1}{R} = \frac{1}{r_1} + \frac{1}{r_2} + \frac{1}{r_3} + \frac{1}{r_4}$$

   If a 1.7-ohm, a 1.48-ohm, a 2.37-ohm, and a 3.35-ohm resistance are connected in parallel, what is the current through the whole system if the voltage is 110 volts?

15. The distance that a freely falling body covers in t seconds is given by the formula $s = 4.9t^2$. How long does it take for an object to fall 363.7 meters?

16. Find the hypotenuse of a right triangle having legs of 137.8 and 289.6.

17. The volume of a sphere is found by the formula $V = (4\pi r^3)/3$, where r is the radius of the sphere. A large solid steel sphere having a diameter of 13.87 meters has a spherical core removed. This leaves a hollow core having a diameter of 8.835 meters. What is the volume of the steel remaining?

18. If light travels at the speed of 298 980 000 meters per second, then the distance to the nearest star can be considered to be 4.23 light years. That is, it would take 4.23 years traveling at the speed of light to reach this nearest star. Our astronauts have been traveling at a speed of approximately 40 000 000 meters per hour. At this speed, how many years would it take to reach this star?

19. The volume V of a cone is found by using the formula $V = (1/3)\pi r^2 h$, where r is the radius of the base of the cone and h is the height. A cone is set on its base and filled to half its height with water. If the diameter of the base is 18.4 meters and the height of the cone is 12.7 meters, what is the volume of the water in the cone, in cubic meters.

20. For a regular pentagon with sides of length 1, what is the length of the segment from the center of the pentagon to the midpoint of one of the sides? (This is called the "apothem.")

(Answers on page 114)

JUNIOR–SENIOR LEVEL

Directions: Give your best estimate for the answer to each problem. No calculators or tables are permitted. Bring only a pencil; scratch paper will be furnished. Units of measure are not required on answers. All answers must be expressed to three significant digits using scientific notation, $a \times 10^n$ where $1.0 \leq a < 10.0$ and n is an integer. If the answer is $a \times 10^0$, the 10^0 factor may be omitted.

1. Evaluate: $\dfrac{\sqrt{3} - \sqrt{2}}{\sqrt{2}}$

2. Find the solution to the following equation that is between 1 and 2. Find the answer to three decimal places.
$$x^3 - x - 3 = 0$$

3. Evaluate: $(2.16)^{21}$

4. Let $y = x^3 - 1$. If $|x - 2| < .001$, then $|y - 7| < m$. What is m?

5. Evaluate: $[\log (100\ 000)^{3/2}]^{1/2}$

6. Evaluate: $\dfrac{353(\cos 30°)}{268}$

7. Let $y = x^3 + 3$. If $|y - 11| < .001$, then $|x - 2| < m$. What is m?

8. What is the length of the diagonal of a rectangular solid that is 60 by 15.5 by 54.5?

9. The following is an infinite geometric series for which the terms get increasingly smaller, eventually having little effect on the sum. What would be the sum?

$$\frac{3}{7} + \frac{3}{7}\left(\frac{5}{11}\right) + \frac{3}{7}\left(\frac{5}{11}\right)^2 + \cdots$$

10. The police use the formula $s = \sqrt{30fd}$ to estimate the speed in miles per hour (s) of a car if it skidded a distance (d) in feet. The variable f is the coefficient of friction determined by the type of road and conditions. What is the estimated speed of the car if $f = 0.43$ and $d = 78$ feet?

11. A chord of length 8.3 is in a circle of radius 15.6. What is the distance of the chord from the center of the circle?

12. Suppose the formula for power generated by a windmill is $P = .0143(V^3)$. How fast must the wind be blowing to produce 120 watts of power?

13. Solve for the largest value of x:

$$\begin{vmatrix} x & 3 \\ 9 & x-7 \end{vmatrix} = 0$$

14. Find the measure of $\angle A$ in the following triangle, given the following:

$$m < B = 60°,\ BC = 3.6,\ AC = 4.33$$

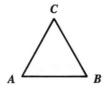

15. Evaluate: $\dfrac{47.5}{\tan (75°)}$

16. Evaluate: $\sqrt{\dfrac{6^2 + 7^2 + 9^2 + 10^2 - \dfrac{(6 + 7 + 9 + 10)^2}{4}}{3}}$

17. Evaluate: $-\begin{vmatrix} \sin{(15°)} & \cos{(15°)} \\ \cos{(15°)} & \sin{(15°)} \end{vmatrix}$

18. Evaluate: $\dfrac{1 - e^{3.5}}{1 + e^{3.5}}$

19. Solve for x:
$$2.4^x + 3.85 = 7.70$$

20. Find the length of AB, if $AC = 5.1$, $BC = 7.9$, and $m\angle C = 45°$.

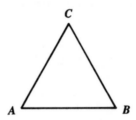

(Answers on page 114)

Multiple-Choice Competition

This is the last written format that will be described. It can be used with any organizational structure. The difference between multiple-choice problems and ordinary written problems is in the nature of the problems themselves and in the speed and accuracy for grading.

When a multiple-choice problem is written, the difficulty for the question writer is in preparing reasonable alternative incorrect choices. Not only must the correct answer be included, but the other choices should be such that a contestant cannot eliminate them simply because they are unreasonable. Furthermore, a contestant can frequently determine the correct answer by trying the answer rather than solving the problem directly.

Multiple-choice problems can be used to evaluate conceptual understanding more easily than open-ended problems, and this will be illustrated in the sample problems. The following problems are open to any students.

1. The smallest number that 12 can be multiplied by to give a perfect cube is
   (A) 144   (B) 1728   (C) 27   (D) 18   (E) 64.

2. The smaller angle between the hands of a clock at 12:25 is
   (A) 132.5°   (B) 137.5°   (C) 150°   (D) 137°   (E) none of these.

3. When 14.2126 is divided by 0.000358, what is the quotient?
   (A) 39.7   (B) 397   (C) 3970   (D) 0.000 397 (E) 39 700

4. In the figure below, a square is divided into a 100 × 100 grid of smaller squares. If 100 squares are shaded in the top row, 99 in the next row, 98 in the third, and so on, what portion of the squares are shaded?
   (A) 1/2   (B) 51/100   (C) 101/200   (D) 201/400   (E) 9/16

5. Which of the following operations with whole numbers will always give a whole number?
   a) addition
   b) multiplication
   c) division
   (A) a only   (B) b only   (C) c only   (D) a and b   (E) b and c

6. If five cats catch six mice in six minutes, how many mice would be caught by thirty cats in thirty minutes?
   (A) 30   (B) 180   (C) 150   (D) 6   (E) 900

7. If ($\sqrt{2}$ − 1) is one root of the equation $3x^2 - 2ax + 1 = 0$, what is the value of a?
   (A) $5\sqrt{2} - 5$ (B) $8\sqrt{2} - 11$ (C) $5\sqrt{2} + 5$ (D) $2\sqrt{2} - 1$ (E) $2\sqrt{2} + 2$

8. A rectangular tank with a square base of side 40 cm contains water to a height of 30 cm. When a solid cube of side 20 cm is totally submerged in the tank, the level of the water rises to what height (in cm)?
   (A) 20   (B) 35   (C) 37.5   (D) 40   (E) 50

9. If x is positive, which of the following expressions must be less than 1?
   (A) $1/x$ (B) $(1 + x)/x$ (C) x^2 (D) $(1 - x)/x$ (E) $x/(x + 1)$

10. A class of 20 students averaged 66% on an exam, and another class of 30 averaged 56% on the same exam. The average percentage for all 50 students was:

    (A) 58   (B) 59   (C) 60   (D) 61   (E) 62

11. Equilateral triangle ABC has an area of $\sqrt{3}$. Point P is an arbitrary point in the interior of the triangle. What is the sum of the distances from P to $\overline{AB}$, $\overline{AC}$, and $\overline{BC}$?

    (A) 1   (B) $\sqrt{3}$ (C) 1/2   (D) $\sqrt{3}/2$ (E) $\sqrt{3}/3$

12. The volume of one cube is twice the volume of another. How many times longer is the edge of the larger cube than the edge of the smaller one?

    (A) 2   (B) $\sqrt{2}$ (C) 8   (D) $\sqrt[3]{2}$ (E) $\sqrt[3]{8}$

13. The product of Mr. Smith's age and the ages of his two children is 4018. How old was Mr. Smith when the first child was born?

    (A) 27   (B) 28   (C) 34   (D) 35   (E) 41

14. A car has an odometer reading of 15 951 miles, which is a palindrome (the number reads the same forward and backward). After two hours of continuous driving at a constant speed, the number is again a palindrome. How fast was the car being driven during those two hours?

    (A) 24.5   (B) 32.5   (C) 50   (D) 55   (E) 60

15. If $x < -4$, what does $|2 - |2 + x||$ equal?

    (A) -4 (B) $4 + x$ (C) $-4 - x$ (D) x (E) $-x$

16. One point on the opposite side of the line $3x + y - 10 = 0$ from the point $(2, 3)$ is

    (A) $(0, 0)$ (B) $(3, 1)$ (C) $(2, 5)$ (D) $(2, 4)$ (E) none of these.

17. Two trains are each traveling toward each other at 90 miles per hour. A passenger in one train notices that it takes five seconds for the other train to pass him. How long, in feet, is the second train? (There are 5280 feet in a mile.)

    (A) 528   (B) 2730   (C) 264   (D) 1320   (E) 240

18. If p, q, and r are each positive integers and $(1/p) + (1/q) = r$, what is the number of possible values for r?

    (A) 1   (B) 2   (C) 3   (D) 4   (E) 5

19. What is the value of $\dfrac{9^{2n} - 3^n}{3^n}$?

    (A) 9^{2n} (B) $3^{2n} - 1$ (C) $9^{2n} - 1$ (D) 26   (E) $3^{3n} - 1$

20. At what latitude is the distance around the earth half of what it is at the equator?

    (A) 30°   (B) 45°   (C) 60°   (D) 90°   (E) 120°

21. A sequence of numbers is of the form $a, a + d, a + 2d, a + 3d, \ldots$. If the kth term of this sequence is equal to m and the mth term is equal to k, what is the nth term equal to?
    (A) $k + m - n$ (B) $n - k - m$ (C) $n - k + m$
    (D) n (E) $k - m - n$

22. In the following, P is the statement "$x \leq 0$," Q is the statement "$x^2 > x$," and R is the statement "$x^2 > 1$." Consider the six propositions below:
    a) If P is true, then R is true.
    b) If Q is true, then P is true.
    c) If R is true, then Q is true.
    d) If P is true, then Q is true.
    e) If Q is true, then R is true.
    f) If R is true, then P is true.
    Which of the propositions above is/are true?
    (A) d only   (B) c only   (C) e only   (D) c and d (E) a and b

23. In the equation $x^2 - 41x + 557 = 0$, what is the sum of the cubes of the roots of the equation?
    (A) 41   (B) 410   (C) 516   (D) 820   (E) 116

24. The sum of the squares of eight consecutive integers is 28 364. What is the minimum integer in the set?
    (A) -63 (B) -56 (C) 56   (D) 63   (E) -59

25. If $\sqrt{x - 1} = \sqrt{x + 1} - 1$, what is $2x$?
    (A) 2.5   (B) $2i$ (C) 0   (D) 5/4   (E) 5/8

26. The points that satisfy the system $x + y = 1$ and $x^2 + y^2 < 25$ constitute the following set:
    (A) only two points   (B) an arc of a circle   (C) a single point
    (D) a line segment excluding the endpoints
    (E) a line segment including the endpoints

27. The arithmetic mean of two numbers is 6 and the geometric mean is 12. What is the quadratic equation that has these two numbers as its roots?
    (A) $x^2 + 12x + 144$ (B) $x^2 - 12x + 144$ (C) $x^2 - 6x + 24$
    (D) $x^2 + 6x + 36$ (E) $x^2 - 6x + 36$

28. The function f is defined recursively as follows:
    $$f(n + 1) = \frac{3[f(n)] + 1}{3}, \text{ for } n = 1, 2, 3, \ldots \text{ and}$$
    $$f(1) = 7.$$
    Find $f(100)$.
    (A) 33   (B) 40   (C) 43   (D) 304   (E) 307

29. The first three terms of a geometric sequence are $\sqrt{5}$, $\sqrt[3]{5}$, $\sqrt[6]{5}$.

    (A) $\sqrt[8]{5}$ (B) $\sqrt[9]{5}$ (C) $\sqrt[10]{5}$ (D) $\sqrt[12]{5}$ (E) 1

30. The points A and B lie on the lines $y = 2x$ and $y = 0$, respectively. If the midpoint of the interval AB is $P(p, q)$ and the length of AB is $2\sqrt{5}$, then which of the following is correct?
    (A) $p^2 + q^2 = 5$ (B) $p^2 + q^2 = 20$ (C) $3p^2 + q^2 = 4(5 + pq)$
    (D) $4p^2 + 5q^2 = 4(5 + pq)$ (E) $p^2 + 2q^2 = 5 + 2pq$

31. Lines can be drawn through the point $(-4, 3)$ so that the sum of their x and y intercepts equals twice the slope of the line. What is the sum of the slopes of all such lines?

    (A) 0   (B) -1 (C) 1/2   (D) $-4/3$ (E) 2

32. In the expansion of $\left(a - \dfrac{1}{\sqrt{a}}\right)^7$, what is the coefficient of $a^{-1/2}$?

    (A) -7 (B) 7   (C) 21   (D) -21 (E) 35

33. The set of points satisfying the pair of inequalities $y > 2x$ and $y > 4 - x$ is contained entirely in which of the following quadrants?
    (A) I and II   (B) II and III   (C) I and III   (D) III and IV (E) I and IV

34. Which of the following is a factor of $x^2 - y^2 - z^2 + x + y - z + 2yz$?
    (A) $(-x + y + z)$ (B) $(x - y - z)$ (C) $(x + y + z)$
    (D) $(x - y + z + 1)$ (E) $(x - y - z + 1)$

35. Given segment AB and a point C that is not on the line AB. As point C moves in a line parallel to the line AB, what is the locus of points determined by the intersection of the three medians of the triangles?
    (A) a line that intersects C
    (B) two lines that intersect A or B
    (C) a line parallel to AB, 2/3 of the way from C
    (D) a line parallel to AB, 1/2 of the way from C
    (E) none of the lines listed above

36. What is the number of solutions in positive integers of $2x + 3y = 763$?
    (A) 255   (B) 254   (C) 128   (D) 127   (E) 0

37. Solve for x: $2^{2x} + 4(2^x) = 1$

    (A) $\log \sqrt{5}$
    (B) $\log (4 - \sqrt{5})$ (D)$\log (\sqrt{5} - 2) - \log 2$

    (C) $\log \left(\dfrac{\sqrt{5} - 2}{2}\right)$ (E) $\dfrac{\log (\sqrt{5} - 2)}{\log 2}$

38. In the figure that follows, $\overparen{AB}$ is a parabolic arch. M and C are the midpoints of $\overline{AB}$ and $\overparen{AB}$. $AB = 40$, $MC = 16$, and $MX = 5$. Find XY.
   (A) 14   (B) 15   (C) 46/3   (D) 31/2   (E) 63/4

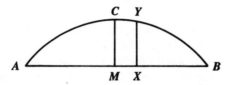

39. If the domain of x is any point such that $x^2 - 6x + 8 < 0$ and $f(x) = x^2 + 6x + 8$, which of the following is correct?
   (A) $2 < f(x) < 4$
   (B) $24 < f(x) < 48$
   (C) $f(x) > 48$
   (D) $0 < f(x) < 48$
   (E) $0 < f(x) < 24$

40. If $0 < x < 1$, $y = x^x$, and $z = x^y$, what are the three numbers arranged in order of increasing magnitude?
   (A) x, y, z   (B) x, z, y   (C) y, z, x   (D) z, y, x   (E) z, x, y

(Answers on page 115)

Oral Competition

   Although it is possible to have an oral presentation on many topics in mathematics, this format seems to be most effective when the contestants are to make a presentation on a sophisticated mathematics topic. The contestant is given a specific reference from which the questions will be determined. For example, the reference might be chapter 4, "Gaussian Integers," in *Enrichment Mathematics for High School,* Twenty-eighth Yearbook of the National Council of Teachers of Mathematics. In this chapter, complex numbers of the form $a + bi$, where a and b are integers, are studied with respect to their characteristics in a number theory sense. In the examples that follow, the specific reference sources will be given. Although the reference sources are known by the contestants, they often have to use additional resources to develop a full understanding of the topic.

   At the competition the contestant will be given one or more problems dealing with the material in the reference that he or she has studied. The problems will be given to the contestant ten minutes prior to the presentation time. During the next ten minutes the contestant may use any resource he or she chooses to prepare for the presentation. Usually, the contestant is allowed to bring only one prescribed sheet of paper into the presentation room.

On entering the presentation room, the contestant should identify himself or herself and his or her school. The contestant then has a maximum of seven minutes of uninterrupted time to give a presentation in response to the problems given. Usually, there is an official timer who gives a warning at five minutes or six minutes into the presentation. At the end of seven minutes the contestant should stop immediately. At this time, judges in the room may ask questions. These may be questions about the contestant's presentation that will help clarify some possible ambiguous information. There may be predetermined questions related to the topic that are asked of all contestants. This may be to determine how a contestant "thinks on his or her feet." Judges may also decide simply to let a contestant continue his or her presentation during the last three minutes.

The presentation is evaluated by two judges, who independently award up to 25 points on the basis of the following criteria:

0–10 Points—*Presentation Skills:* How well organized is the presentation? Are important ideas clear and emphasized? Is the material presented in a logical sequence? Does the contestant speak clearly, and is the visual information written carefully?

0–10 Points—*Knowledge of Subject:* Does the contestant have correct solutions to the questions posed? Are there mistakes in the work presented? Is the logic correct? Is the correct notation used? Does the contestant demonstrate a firm understanding of the material by citing relevant theorems, axioms, and so on, when appropriate?

0–5 Points—*Spontaneous Response Skill:* How well does the contestant respond to the questions posed by the judges during the final three minutes? (If there is no discussion, these 5 points can be included in the previous category.)

Because judging oral presentations is primarily a subjective task, the same judges, if possible, should evaluate all presentations for a given topic. If there are more than six schools in a contest, this becomes quite difficult. Each presentation is ten minutes, and there must be time between presentations for a judge to assign points and to make comments. (It is important to the contestants that they receive written comments on their presentation.) Therefore, it takes about two hours to have six oral presentations.

If there are more than six presentations, it is suggested that more than one room be used, but that the judges have an orientation session and be rotated during the competition. For example, suppose there are to be twelve presentations in Rooms I and II. The following chart is a possible rotation scheme for the judges:

Room I		Room II	
Contestant	Judges	Contestant	Judges
1	A,B	7	C,D
2	A,B	8	C,D
3	A,C	9	B,D
4	A,C	10	B,D
5	A,D	11	B,C
6	A,D	12	B,C

Using this scheme, any two judges are together for only two presentations. If, preceding the contest, there is an orientation session in which the judges discuss the merits or demerits that may arise in a presentation and the procedures for assigning points, scoring will be more consistent than it otherwise would be.

As you will notice in the examples that follow, there are typically three problems for each topic. The first problem is usually general and allows the contestant to indicate at least that he or she understands the material in the reference. The second problem may be one that was posed in the reference. The third problem can be a moderate extension of the reference material. The "Judges' Information" sections are not intended to present well-articulated solutions to the problems but are merely suggestions that a judge may use in conjunction with the reference when evaluating a student's presentation. In the examples that follow, possible additional questions for the judges to ask are given. A central committee may decide, however, that there should be no predetermined questions and that judges should ask questions related only to the contestant's main presentation.

TOPIC: THE CONCEPT OF NUMBER

Reference: Niven, Ivan. "The Concept of Number." In *Insights into Modern Mathematics*, Twenty-third Yearbook of the National Council of Teachers of Mathematics, pp. 7–24. Washington, D.C.: 1957.

Problems

1. Define and explain the following:
    a) Rational numbers and irrational numbers
    b) Algebraic numbers and transcendental numbers

2. Give examples of a real irrational algebraic number and a complex irrational algebraic number, and explain why they fall into these classifications.

3. Prove that if α is a rational number and β is a transcendental number, then $\alpha\beta$ is a transcendental number.

Judges' information

1. The definitions are given in the text—on pages 12 and 20, respectively—and should not be too difficult. However, the student may forget the $b \neq 0$ restriction for the rationals, or that all coefficients must be *integers* for the definition of algebraic numbers.

2. There may be many examples, but if the justifications are to be brief, they should probably select $\sqrt{2}$ and $i\sqrt{2}$. The proof of their irrationality is in the text and the polynomials necessary for their being algebraic are $x^2 - 2$ and $x^2 + 2$, respectively.

3. Briefly, the proof might go as follows:
   Assume $\alpha\beta$ is an algebraic number and $\alpha = p/q$, with $q \neq 0$. Therefore, there are integers $a_n, \ldots, a_1, a_0$ such that
   $$a_n\left(\frac{p}{q}\beta\right)^n + \ldots + a_1\left(\frac{p}{q}\beta\right) + a_0 = 0.$$

   Multiplying through by q^n you obtain
   $$(a_n p^n)\beta^n + \ldots + a_1 pq^{n-1}\beta + a_0 q^n = 0.$$
   Since all the coefficients are integers, $\alpha\beta$ is algebraic because it is a solution to the polynomial
   $$(a_n p^n)x^n + \ldots + (a_1 pq^{n-1})x + a_0 q^n = P(x).$$
   This is a contradiction, and therefore $\alpha\beta$ is not algebraic; that is, it is transcendental.

Additional questions

1. If two numbers are irrational, is their product irrational? Explain.
   Answer: No, a simple example is $\sqrt{2} \cdot \sqrt{2} = 2$.

2. Show that algebraic numbers can be rational or irrational.
   Answer: A linear equation, $ax + b = c$, has a rational solution, and many quadratic equations have irrational solutions.

TOPIC: DIOPHANTINE EQUATIONS

Reference: Stewart, B. M. *Theory of Numbers*, pp. 96–103. Riverside, N.J.: Macmillan, 1964.

Problems

1. Define and explain the following expressions:
   a) Diophantine equation
   b) System of Diophantine equations

2. Given a quadratic Diophantine equation in one variable, explain why a necessary condition for a solution is that the discriminant be a perfect square. Is this a sufficient condition?

3. Find all solutions to the Diophantine equation $x^2 - 4y^2 = 8$. Describe the steps taken to find the solutions.

4. Given the Diophantine equation $x^2 - 4y^2 = k$, find several values of k for which there are solutions. Explain how you selected these values.

Judges' information

1. a) A Diophantine equation has solutions that are restricted to integers (often positive integers).

   b) If the solution to a system is restricted to n-tuples of integers (lattice points), the system is said to be a Diophantine system.

2. For any solution to a quadratic to be *rational*, the discriminant must be a perfect square. Therefore, this is a necessary condition. However, it is not a sufficient condition. Suppose $s^2 = b^2 - 4ac$, then the solution is integral if $-b \pm x \equiv 0 \pmod{2a}$.

3. The following is a brief possible solution. (There are *no* integral solutions.)
   $x^2 - 4y^2 = (x + 2y)(x - 2y) = 8 \Rightarrow$ for solutions to exist there must be integers r, s such that $rs = 8$, $x + 2y = r$, $x - 2y = s$

   $$\left. \begin{array}{l} x + 2y = r \\ x - 2y = s \end{array} \right\} \Rightarrow 2x = r + s \text{ and } 4y = r - s$$

   Therefore, r and s must have the same parity. That is, they are both even or both odd. Hence, $r \cdot s = 8 \Rightarrow r = \pm 2; s = \pm 4$. They cannot be the ± 1 and ± 8 factors. Hence, $2x = \pm 6$ and $4y = \pm 2$. But this implies that y is *not* integral.

   Therefore, there are no solutions.

4. This might be easy—just finding a couple of values, for example, 4, 16, 20. Or this can be very complex—looking for generalizations. (Note whether the students use a process or just trial and error.) Using parity information, as above, we find that the values of k are

    a) multiples of 16
    b) $4(2n + 1)$
    c) $4n + 1$.

Additional questions

1. If a system of equations has only positive integer solutions, is that system necessarily Diophantine? Explain.

   Answer: No, the solutions must be restricted to positive integers. It is the restriction that makes them Diophantine, not the fact that the solutions happen to be integers.

2. Do Diophantine equations have only a finite number of solutions? Explain.

   Answer: Not necessarily. For example, the equation $x - y = 1$ has many positive integer solutions.

FIBONACCI AND LUCAS NUMBERS

Reference: Hoggatt, V. E., Jr. *Fibonacci and Lucas Numbers*, pp. 1–29. Boston: Houghton Mifflin Co., 1969.

Problems

1. Compare the Fibonacci and Lucas sequences. Explain how other such sequences can be generated. Write the first five terms of a contest (C) sequence where $c_1 = 2$ and $c_2 = 4$. What characteristics does it have with the Fibonacci and Lucas sequences?

2. Assume the solutions to the equation $x^2 = x + 1$ to be α and β.

   Prove that the *n*th term of a Fibonacci sequence $F_n = \dfrac{\alpha^n - \beta^n}{\alpha - \beta}$.

3. Assuming the *n*th term of the Lucas sequence to be $L_n = \alpha^n + \beta^n$, express C_n in terms of α and β. Show that your expression holds for the first three terms.

Judges' information

1. The first question is very open-ended. You should have little difficulty getting a range of responses, the least of which being that each sequence is formed by adding the two preceding terms and the most of which

being that the ratios of consecutive terms converge to 1.618 and so on. A contestant may note at this point that $C_n = F_n + L_n$. This bit of information will be quite useful in answering the third question.

2. This proof is on pages 10 and 11 in the text. It is reasonably difficult and will take a firm understanding of the content to present the proof in a contest format.

3. This is not a difficult problem, but it is material that is not discussed explicitly in the text. Once the contestant recognizes the relationship $C_n = F_n + L_n$ the algebra is direct: $C_n = (\sqrt{5} + 1) \alpha^n + (1 - \sqrt{5})\beta^n$.

Additional questions

1. Can you describe a golden rectangle and give the ratio of the length to the width?

   Answer: The most geometrically descriptive is the rectangle for which a similar rectangle is formed in a square having sides equal to the width of the original rectangle drawn. For example:

   ABED is a square, and
   ACFD is similar to
   CFEB.

   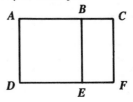

   The ratio is $\dfrac{1 + \sqrt{5}}{2} \approx 1.61$.

2. Consider the ratios of consecutive terms of a Fibonacci sequence $\dfrac{1}{1}, \dfrac{2}{1}, \dfrac{3}{2}, \dfrac{5}{3}, \dfrac{8}{5}, \dfrac{13}{8}, \ldots$ What can you say about them?

   Answer: The ratios approach $\dfrac{1 + \sqrt{5}}{2} \approx 1.61$.

GAUSSIAN INTEGERS

Reference: Calloway, Jean M. "Gaussian Integers." In *Enrichment Mathematics for High School*, Twenty-eighth Yearbook of the National Council of Teachers of Mathematics, pp. 46–55. Washington, D.C.: NCTM, 1963.

Problems

1. Consider the set of Gaussian integers. Do all elements of the set other than 0 have a multiplicative inverse in the set? Explain.

2. Gaussian integers can be defined in two ways:
   a) Complex numbers of the form $a + bi$, where a and b are rational integers
   b) Complex numbers of the form $a + bi$, which are solutions to a quadratic equation $x^2 + mx + n = 0$, where a and b are rational, and m and n are rational integers

   Show that these two definitions are equivalent.

3. If the cubic expression $x^3 + ax^2 + bx + 91$ can be factored to the expression $(x - p)(x - q)(x - s)$, where p, q, and s are Gaussian integers, what are the possible sets of values for p, q, and s?

Judges' information

1. To show that many Gaussian integers have no inverse in the set, it is assumed that the student will use the notion that a Gaussian integer is of the form $a + bi$, where a and b are rational integers. All they need to demonstrate is that the multiplicative inverses will "usually" have non-integral real and imaginary parts. Even one counterexample is sufficient.

2. To show equivalence, the student must demonstrate that if *either* is accepted, then the other can be shown as a theorem. Accepting "a" as a definition and showing that "b" is true is fairly simple. If one complex number, $a + bi$, is a Gaussian integer, then use its complex conjugate, $a - bi$. The two then are the solutions of the quadratic equation $x^2 + 2ax + (a^2 + b^2)$. Because a and b are integers, $2a$ and $a^2 + b^2$ are also rational integers. Going the other direction is more difficult, but it is done in the text. In doing the proof the student will eventually reach a point where $a = m/2$ and $b = c/2$, and $m^2 + c^2 = 4q$. The equation $m^2 + c^2 = 4q$ means that m and c are both even or both odd. If time is of the essence and it is apparent that the student can do the exercise, a judge may suggest that the student omit the proof. However, if a student does do the proof, it should not be too difficult to examine the cases for $(2m + 1)^2 + (2n + 1)^2$. In any case, it should be argued that m and c are both even; in which case a and b are rational integers.

3. This question is actually a shorter one than the second, but, whereas the second question is discussed in the text, this question will require the use of several theorems to solve the problem. To find the possible combinations that will work, the contestant will have to factor 91 to 7(13), then use the theorems regarding Gaussian "primeness" of numbers of the form $4n + 3$ (i.e., 7) and the Gaussian "nonprimeness" of numbers of the form $4n + 1$ (i.e., 13). Along with this, students should understand

the concept of a Gaussian unit to form the appropriate sets of possibilities. For example, if one set is 1, 1, and 91, then i, $-i$, and 91 is another possibility. Similarly, one set is 7 multiplied by $7i$, $2 + 3i$, and $2 - 3i$. Multiply this set by $-i$, and it will generate $7i$, $-2i + 3$, and $2 - 3i$.

Additional questions

1. Show that 2 is a Gaussian integer using the second definition.

   Answer: Any quadratic equation with $(x - 2)$ as one factor and $(x - a)$ as the other factor. For example, $x^2 - 4 = 0$.

2. Show the real integer 5 as a product of two Gaussian integers.
   Answer: $5 = (2 + i)(2 - i)$

GROUPS

Reference: Allendoerfer, Carl B., and Cletus O. Oakley. "Groups." In *Principles of Mathematics*, pp. 69–82. New York: McGraw-Hill, 1963.

Problems

1. Consider the following set of four elements representing simple permutations. If * represents the usual permutation "operation," then show that this set forms an Abelian group with respect to the operation *.

$$I = \begin{pmatrix} 1\ 2\ 3\ 4 \\ 1\ 2\ 3\ 4 \end{pmatrix} \qquad A = \begin{pmatrix} 1\ 2\ 3\ 4 \\ 1\ 2\ 4\ 3 \end{pmatrix} \qquad B = \begin{pmatrix} 1\ 2\ 3\ 4 \\ 2\ 1\ 3\ 4 \end{pmatrix} \qquad C = \begin{pmatrix} 1\ 2\ 3\ 4 \\ 2\ 1\ 4\ 3 \end{pmatrix}$$

2. Is the set above isomorphic to the set of integers modulo 4 with respect to addition? Explain.

3. Consider the set of congruences of a (nonsquare) rectangle with itself. That is, if $PQRS$ is a rectangle, then the four congruences might be identified as

   I: a rotation of 360° x: a rotation of 180°
   y: a reflection over a line perpen- z: a reflection over a line perpen-
      dicular to one pair of sides   dicular to and bisecting the
      (the bisector)   other pair of sides.

   Is the original group isomorphic to this group if this group is defined with respect to the usual composition operation? Explain.

Judges' information

1. Based on the information in the reference, the student will probably set up a table to demonstrate the group relations. It should look something like this:

$*$	I	A	B	C
I	I	A	B	C
A	A	I	C	B
B	B	C	I	A
C	C	B	A	I

. This is an Abelian group, that is, closed, associative, commutative, has an identity (I), and each element has itself as an inverse.

This group is not isomorphic to the integers modulo 4 under addition, but it is isomorphic to the congruences of a rectangle with itself.

To show the second case they must note that the operations preserve the one-to-one correspondence.

Note: The table for the group of rotations has the following table:

$*$	i	x	y	z
i	i	x	y	z
x	x	i	z	y
y	y	z	i	x
z	z	y	x	i

. So, the mapping is $I \leftrightarrow i$, $A \leftrightarrow x$, $B \leftrightarrow y$, $C \leftrightarrow z$.

Additional questions

1. Are the integers a group with respect to ordinary multiplication? If not, why not?

   Answer: No, because to have inverses, the set must have rational numbers.

2. If $A = \begin{pmatrix} 1 & 2 & 3 & 4 & 5 \\ 2 & 3 & 5 & 1 & 4 \end{pmatrix}$, then what is the inverse of A?

   Answer: $A^{-1} = \begin{pmatrix} 1 & 2 & 3 & 4 & 5 \\ 4 & 1 & 2 & 5 & 3 \end{pmatrix}$

INEQUALITIES

Reference: Beckenbach, E. F., and Bellman, R. *An Introduction to Inequalities*, New Mathematical Library, no. 3, pp. 79–98. Washington, D.C.: Mathematical Association of America, 1975.

Problems

1. When Dido's problem is applied to rectangles, it can be shown that for a fixed perimeter, a square generates the maximum area; and the reverse of this problem is that for a fixed area, a square generates the minimum perimeter. State the problem and its reverse as applied to
   a) triangles and
   b) rectangular solids.

2. Consider the following statement: The sum of the edges of a rectangular solid is k, and the surface area is m^2. Prove: $2m \sqrt{6} \leq k$.

3. If the points (a, b), $(-a, b)$, $(-a, -b)$, $(a, -b)$ represent the vertices of a rectangle, find the ellipse of minimum area that circumscribes this rectangle.

Judges' information

1. a) The problem applied to triangles is obvious, using "equilateral triangle" rather than "square."
   b) For rectangular solids, the problems become the relationship between the *surface area* and *volume* of a cube.

2. This is a slight adaptation of an exercise in the text. Given the two basic formulas
$$2(lw + lh + wh) = m^2$$
$$4l + 4w + 4h = k,$$
then since $(l + w)^2 \geq 0$, $l^2 + 2lw + w^2 \geq 0$. Therefore, $2lw \leq l^2 + w^2$. Similarly, $2wh \leq w^2 + h^2$ and $2hl \leq h^2 + l^2$
$\Rightarrow lw + wh + hl \leq l^2 + w^2 + h^2$
$\Rightarrow 3(lw + wh + hl) \leq l^2 + w^2 + h^2 + 2lw + 2wh + 2hl$
$\Rightarrow 3(lw + wh + hl) \leq (l + w + h)^2$
$\Rightarrow 3\left(\dfrac{m^2}{2}\right) \leq \left(\dfrac{k}{4}\right)^2$
$\Rightarrow 3m^2 \leq \dfrac{k^2}{8}$
$\Rightarrow 24m^2 \leq k^2$.

3. This is similar to the wealthy football player problem, which is probably too time consuming. To do the problem, the student will probably use the reverse, that is, finding a rectangle of maximum area inscribed in a given ellipse. Assume this ellipse is defined by

$$\frac{x^2}{A^2} + \frac{y^2}{B^2} = 1.$$

Thus given the A.M.–G.M. relationship, for the two expressions

$\frac{x^2}{A^2}$ and $\frac{y^2}{B^2}$, $\left[\left(\frac{x^2}{A^2}\right)\left(\frac{y^2}{B^2}\right)\right]^{1/2} \leq \frac{1}{2}\left(\frac{x^2}{A^2} + \frac{y^2}{B^2}\right) \Rightarrow \frac{xy}{AB} \leq \frac{1}{2}\left(\frac{x^2}{A^2} + \frac{y^2}{B^2}\right)$ where

x, y are points on the ellipse, so

$$\frac{x^2}{A^2} + \frac{y^2}{B^2} = 1.$$

Therefore $xy/AB \leq 1/2$. But the equality, or *minimum* value, holds only when

$$\frac{xy}{AB} = \frac{1}{2}\left(\frac{x^2}{A} + \frac{y^2}{B}\right),$$

which implies

$$\frac{x^2}{A^2} - \frac{2xy}{AB} + \frac{y^2}{B} = 0;$$

that is,

$$\frac{x}{A} = \frac{y}{B}.$$

Therefore, the inequality/equality is a maximum if $\frac{x^2}{A^2} = \frac{y^2}{B^2} = 1/2$.

That is, if $x = \frac{A}{\sqrt{2}}$, $y = \frac{B}{\sqrt{2}}$.

However, using the reverse, which is the original problem, we note that the rectangle has coordinates $(|a|, |b|)$, or $|a| = \frac{A}{\sqrt{2}}$ and $|b| = \frac{B}{\sqrt{2}}$; thus $A^2 = 2a^2$ and $B^2 = 2b^2$. Therefore, the ellipse becomes

$$\frac{x^2}{2a^2} + \frac{y^2}{2b^2} = 1.$$

Additional questions

1. If $a < b$ and $c < d$, is $ac < bd$? Explain.
   Answer: No, because the numbers must be positive.

2. What are the relationships between polygons with various numbers of sides having the same perimeter, and the areas of these polygons? Explain, but don't prove your answer.
   Answer: The more sides, the greater the area. The circle is the maximum.

INVERSIONS OF CIRCLES

Reference: Smart, J. R. *Modern Geometrics.* 2d ed., pp. 313–39. Monterey, Calif.: Brooks/Cole Publishing Co., 1978.

Problems

1. Suppose $\triangle ABC$ is a right triangle inscribed in a circle. The center of the circle, O, is on the hypotenuse, $\overline{AC}$. Describe the inversions of the figure with respect to the given circle. Describe the inversions of the figure with respect to other circles that are concentric with the given circle.

2. With respect to the same figure as the one used above, describe the inversions of the figure using the following points as centers and the radii listed. In each case tell which arcs are orthogonal.

   a) B is the center of inversion and $\overline{BC}$ is the radius.

   b) The center is in the exterior of the figure, and the entire figure is in the interior of the circle of inversion and not on the lines of any side of the triangle.

   c) The center is on $\overset{\frown}{AC}$ and the circle of inversion is tangent to $\overline{AC}$.

3. Under inversion with respect to a sphere, describe the inversion of two intersecting planes. Include in your discussion a discussion of the inversion of the line of intersection.

Judges' information

1. In the inversion of the figure, the circle will remain invariant, the hypotenuse $\overset{\frown}{AC}$ will become a line $\overline{AC}$, and the sides of the triangle will be "petals," that is, arcs of circles, having one endpoint at B and the other endpoint at the appropriate end of the diameter. These petals will be in the exterior of the circle of inversion. For concentric circles of inversion the figure will be similar, except for the size.

2. Rather than describing each of the figures, it might be easier to note some of the important transformations for each of the inversions:

   a) The inversion of $\overline{BC}$ will be a ray with endpoint on the circle of inversion. $\overset{\frown}{BA}$ will be transformed to a ray with endpoint inside or outside depending on the length of $\overline{BA}$. The circle will be inverted to a circle tangent to the circle of inversion. AC will be an arc of a circle.

   b) In this case each of the sides becomes an arc of the circle in the interior of the circle of inversion, and the circle is inverted into another circle. In particular, the arcs AB and BC are orthogonal, and AC is orthogonal to the inverted circle. This relationship between AC and the inverted circle is the relationship used in the Poincaré model for hyperbolic geometry.

c) The two important characteristics to note here are that the inversion of the circle is now a line, and the inversion of the hypotenuse is a circle that is internally tangent to the circle of inversion. The inversion of the circle must be orthogonal to the inversion of the hypotenuse, so it must be a diameter of the circle.

3. This may not be a difficult part, but it is assumed that the inversions in three space are more difficult to conceive than inversions in a plane. The inversions of planes will be spheres, and therefore the inversions of intersecting planes are intersecting spheres. (An exception is the case where the plane contains the center of the sphere, in which case the inversion is in the same plane.) It should be noted that the inversion of the intersection is a circle and that this circle is in the same plane as the line.

Additional questions

1. If you know that the inversion of a circle is a line, what do you know about the center of inversion?
   Answer: The center of inversion must be on the circle.

2. If one sphere contains the center of a sphere of inversion and intersects the sphere in a circle, what is the inversion image of that sphere and where is it?
   Answer: The image is a plane that is determined by the circle of inter-
   section.

LAWS OF GROWTH AND DECAY (Calculators allowed)

Reference: Leithold, Louis. *Calculus and Analytic Geometry*, pp. 420–27.
          New York: Harper and Row, 1976.

Problems

1. Explain what is meant by the laws of natural growth and natural decay. If each week a company increases its profits by $100, is this an example of natural growth? Explain.

2. A certain bacteria doubles every 10 seconds. If you assume that the rate of growth is constant, by what ratio would the bacteria grow in 20 seconds? By what ratio would it grow in 25 seconds? Explain your solutions.

3. The Simplex Bank offers an 8% savings account that they will compound quarterly. The Complex Bank, however, states that it compounds continuously. If you put $1000 in both savings accounts, each at 8%, then how much more could you expect to have in the Complex Bank at the end of one year? What percentage would the Complex Bank have to offer to match the amount offered by the Simplex Bank?

Judges' information

1. This entire topic is somewhat strange because there is not much conceptual information. To understand the material, the contestant has to understand how to apply the principle to a variety of situations. The first question is asked merely to determine whether a contestant understands the central principle. The question regarding the increase in profits of $100 is not an example of natural growth because the growth has to be by a constant ratio.

2. For this problem, the student has to apply the principle for the "25 second" part of the question. For the "20 second" part, he or she should simply say that the bacteria have quadrupled. For the 25 seconds, the bacteria will have increased by a multiple of 5.657.
   In general, $A = Ce^{kt}$. Let $C = 1$, and for $t = 10$, $A = 2$. Therefore, solving for k,
   $$k = 0.1(\ln 2).$$
   Therefore, $A = e^{25k} = 5.657$.
   It is assumed that the students will be able to determine the answer using a calculator.

3. The first part of the question is reasonably straightforward using the procedure described in the text. The amounts at the end of a year will be $1082.43 in the Simplex Bank and $1083.29 in the Complex Bank—a difference of $0.86. (Big deal!) In the second part, the contestant will have to solve for interest. This is a more difficult mathematical task. If he or she does so correctly, then the interest given by the Complex Bank would be 7.96%.

Additional questions

1. What is the half-life of a substance?
   Answer: It is the time required for a substance to decay to half of its original quantity.

2. If the half-life of a substance is 1 year, how long would it take (approximately) for a substance to decay to about 10% of its original quantity?

Answer: It would be between 3 years $\left(\frac{1}{8}\right)$ and 4 years $\left(\frac{1}{16}\right)$, or about

$3\frac{1}{4}$ years.

MASS POINTS

Reference: Sitomer, Harry, and Conrad, Steve R. "Mass Points." *Eureka* 2
(April 1976): 55–62. A magazine sponsored by the Carleton-
Ottawa Mathematics Association, a chapter of the Ontario Association for Mathematics Education, Ottawa, Ontario, Can-
ada.

Problems

1. Define and explain *mass point,*
    addition of mass points, and *centroid.*

2. Use these concepts to find *AD:AE*
    and *BD:BF. E* is the midpoint of *BC,*
    and *CF:FA* = 1:2. Explain your
    answer.

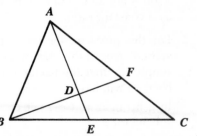

3. State and explain the theorems of Ceva and Menelaus and their con-
    verses.

4. Suppose $\triangle ABC$ has perimeter $2p$, and that $AB + BD = p$, $BC + CE$
    $= p$, and $CA + AF = p$. Prove AD, BE, CF are concurrent. (Hint: Let
    $AB = c$, $BC = a$, $CA = b$. Then determine each of the six segments in
    terms of p, a, b, and c.)

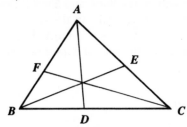

Judges' information

1. The definitions are on the first page. This is merely the concept of a
    moment in physics. This is a "balancing" principle for which the product

of the weight and distance from the centroid must be equal for both points.

2. For the problem, the weight of B and C must be equal (i.e., the "balance point" or fulcrum is in the center) while A must weigh ½ of C. (Heavier objects must be closer to the fulcrum.) Any weight having these ratios would be adequate. 1A, 2B, and 2C are probably the easiest. $2B + 2C = 4E$. Hence, the ratios of the weight of A and D are 1:4, and therefore the lengths are 4:1. Hence, $\overline{AD}{:}\overline{AE} = 4/5$. Similarly, $\overline{BD}{:}\overline{BF} = 3/5$.

3. Both theorems are in the paper. Ceva's theorem refers to concurrency and Menelaus' theorem to collinearity. In both cases, the converse is actually more useful, that is, if ratios hold them concurrent (collinear). Also, be sure the students note the two cases with Menelaus' theorem; that is, when 1 or 3 points are exterior to the triangle.

4. This is a direct application of Ceva's theorem. If $BC = a$, $AC = b$, and $AB = c$, it can be shown that $AF = p - c$, $BD = p - c$, $CE = p - a$, and that $FB = p - a$, $DC = p - b$, $AE = p - c$. Obviously $(AF)\,(BC)\,(CE) = (FB)\,(DC)\,(AE)$. Therefore, the three segments must be concurrent.

Additional questions

1. If a person weighing 50 kg and another weighing 100 kg are standing on a uniform board, where must a pivot be placed so that the two balance? How does this compare with the definitions related to the mass points?

   Answer: The pivot would have to be equivalent to the "sum" of the mass points, using the text definition.

2. What weight would you assign to points A and B if you wanted the centroid to be $\frac{3}{5}$ of the way from A to B?

   Answer: Any ratio of weight such that A is 2 and B is 3.

MATHEMATICAL INDUCTION

Reference: Dolcani, A., et al. *Modern Introductory Analysis*, pp. 69–74. Boston: Houghton Mifflin Co., 1964.

Problems

1. State the principle of mathematical induction, describing it as clearly as possible in your own words.

2. Given an example where the first step in an induction proof is true, but the second step is not true, and an example where the second step is true, but the first step is not true.

3. Prove that for all natural numbers, $2^{2n+1} + 1$ is a multiple of 3.

Judges' information

1. This question was written in as straightforward a manner as possible, being limited to the first six pages of the reference (*Modern Introductory Analysis*). It is assumed that this type of competition will be new to many of the students.

   This question is obvious, the conditions being, "If (1) $1 \in S$, and (2) $k \in S \to k + 1 \in S$, then $S = N$." Differences in performance will probably be due to the quality of presentation.

2. Although this is an open-ended question and may yield some unusual responses, the student can get an example of each by simply reading the text on page 74. The justifications for the claims are not in the book, and the differences in presentation can be used to discriminate among the students.

3. This is not in the text, but it is not very difficult, and if the contestants understand the process, they should be able to do the proof.

   Step 1: $1 \in S$, because $2^{2(1)+1} + 1 = 9$ is a multiple of 3.

   Step 2: To show $k \in S \to k + 1 \in S$, it is necessary to show that if $2^{2(k+1)} + 1$ is a multiple of 3, then $2^{2(k+1)+1} + 1$ is as well.

   $2^{2(k+1)+1} + 1 = 2^2 \, 2^{2k+1} + 1 = (3 + 1) \, 2^{2k+1} + 1 = 3 \cdot 2^{2k+1} + 2^{2k+1} + 1$, which is a multiple of 3.

Additional questions

1. Is this principle appropriate if S is a finite set?
   Answer: Obviously not, because the second step is meaningless.

2. Is it possible to use this principle if the theorem doesn't hold for 1, but does hold for 2 and for $k \to k + 1$?

   Answer: Yes, but there should be some modification of the theorem, usually replacing n by $n + 1$. Then, when n begins at 2, $n + 1$ begins at 1.

NON-EUCLIDEAN GEOMETRY

Reference: Smart, J. R. *Modern Geometrics*, 2d ed., pp. 199–232. Monterey, Calif.: Brooks/Cole Publishing Co., 1978.

Problems

1. State the characteristic postulate of elliptic geometry, and explain how a sphere can be used as a possible model for the system.

2. Prove that in elliptic geometry the summit angles of a Saccheri quadrilateral are congruent and obtuse.

3. Let *ABCD* be a Lambert quadrilateral in elliptic geometry having right angles, ∠*A*, ∠*B*, and ∠*C*. Prove that *AD* is less than *BC*.

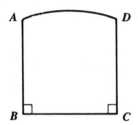

Judges' information

1. The characteristic postulate of elliptic geometry is that any two lines in a plane must meet in an ordinary point. (This contrasts with hyperbolic geometry, where lines meet at an ideal point.) The second half of the question is somewhat open ended. The quality of the response can be used for discrimination.

2. To prove that the upper angles are congruent, the contestant will simply have to draw the diagonals and use Euclidean methods to complete the proof. It should not be too difficult. The following is a possible proof that the upper angles are obtuse. As might be the case with every proof, one may not know specifically where to start. This proof will assume the following theorem: If △*RST* is a right triangle with right angle *R*, then if *RS* is less than a polar distance *q*, then the opposite angle, ∠*T*, is acute. Some of the contestants may prove this as a lemma, but time may prevent them from getting through with it.

   Now suppose *ABCD* is a Saccheri quadrilateral with right angles, *A* and *B*. Therefore, *AD* and *BC* meet at the pole, *P*. The ray *MN* through the midpoints of the two bases is also perpendicular to $\overline{AB}$, and therefore

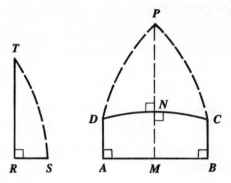

must have P as a pole point. MP must be of length q and NP must be less than q. By the previous theorem $\angle NDP$ is acute and, by doing a little algebra, $\angle ADC$ is obtuse.

3. Here is a possible proof:

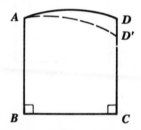

Suppose $AD > BC$. Then there is a D' on AD where $AD' = BC$. Hence, $ABCD'$ is a Saccheri quadrilateral. This means that $m \angle BCD'$ is 90°. This would, of course, lead to the conclusion that $m\angle DCD' < 0$, which is impossible. Therefore, AD cannot be greater than BC.

AD cannot be equal to BC because that would immediately make $ABCD$ a Saccheri quadrilateral. This would then mean that $\angle C$ is obtuse. But it is a right angle, and this is then a contradiction.

Therefore, all that is left is that $AD < BC$.

Additional questions

1. How can we use the expression *the polar distance?* Is it the same for all lines?

   Answer: Once the polar distance is determined for one line, it can be proved that this is the polar distance for all lines in that plane. I would not encourage a participant to prove it, because it is very time consuming.

2. What is the length of a line in elliptic geometry?

Answer: Because the polar distance is constant for all lines in the s plane, the length of a line can be shown to be four times the length of its polar distance. Therefore all lines are four times the length of the polar distance.

NUMBER THEORY

Reference: Egan, Lawrence C. "Number Theory." *Enrichment Mathematics for High School,* Twenty-eighth Yearbook of the National Council of Teachers of Mathematics, pp. 56–64. Washington, D.C.: NCTM, 1963.

Problems

1. Define and explain what an arithmetic sequence is and what a geometric sequence is. Give an example of each, describing the characteristic components, such as common differences or common ratios.

2. State and describe the summation formulas for both arithmetic and geometric sequences. Explain how an infinite geometric sequence can be used to find the rational number equivalent of $1.23\overline{23}$.

3. Can the first three terms of a geometric sequence be the same as the first three terms of an arithmetic sequence? Explain fully your answer, describing circumstances for which the condition may or may not hold.

Judges' information

1. This first part is a straightforward question. Of the various components that should be mentioned, the common difference between the terms of an arithmetic sequence and the common ratio of the terms of a geometric sequence are important. However, a polished presentation may note that the $(n + 1)$th terms would contain a dn for arithmetic sequences and an r^n for geometric sequences. That is, students may choose to elaborate somewhat.

2. The summation formulas for the three sequences are as follows:
   a) $S(\text{arithmetic}) = (n/2)(2a + [n - 1]d)$
   b) $S(\text{geometric}) = (ar^n - a)/(r - 1)$ $\qquad\qquad r \neq 1$
   c) $S(\text{infinite geometric}) = a/(r - 1)$ $\qquad\qquad |r| < 1$
   For the problem, students should use the third formula with $a = 1.23$ and $r = 0.01$. Therefore, $N = S = 1.23/0.99 = 41/33$.

3. This is not a problem in the text but should be a simple extension of the material studied. Probably the simplest way to discuss the problem is to let the first three terms of an arithmetic sequence—a, $a + d$, and $a + 2d$—be a geometric sequence (or let the first three terms of a geometric sequence—a, ar, and ar^2—be the terms of an arithmetic sequence). By algebraic techniques, the presenter should be able to conclude that either $d = 0$ or $r = 1$. Therefore, the only way that the first three terms can be either arithmetic or geometric is if the sequence is one where all terms are the same.

Additional questions

1. Suppose -3, 9, and n are the first three terms of a sequence and $n > 0$. Can the sequence be arithmetic? Geometric? Explain.

   Answer: It cannot be geometric, because n would have to be -27, but it may be arithmetic—though not necessarily.

2. Suppose 4, x, and 9 form three terms of a geometric sequence. What is x?

   Answer: Be sure the presenter indicates that x can be 6 or -6.

POINTS AND LINES CONNECTED WITH A TRIANGLE

Reference: Coxeter, H. S. M., and S. L. Greitzer. *Geometry Revisited*, New Mathematical Library. Washington, D.C.: Mathematical Association of America, 1967. no. 19, pp. 1–26.

Problems

1. Define an orthocenter, centroid, and circumcenter of a triangle. Sketch a diagram using an obtuse triangle describing these points. Use A, B, and C to denote the vertices of the triangle. H should denote the orthocenter; G, the centroid; and O, the circumcenter.

2. For the triangle in part 1 (above), show that H, G, and O are collinear.

3. If R is the radius of the circumscribed circle, prove
$$OH^2 = 9R^2 - AB^2 - AC^2 - BC^2.$$

Judges' information

1. The orthocenter is the intersection of the three altitudes, the centroid is the intersection of the three medians, and the circumcenter is the inter-

section of the three perpendicular bisectors of the sides. In the following diagram, you should note that the orthocenter and the circumcenter are in the exterior of the triangle. (Make sure the contestant's diagram is large.)

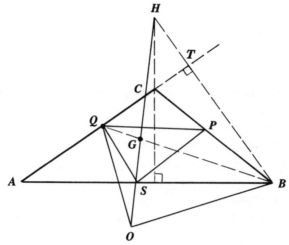

2. This is a proof in the text, using an acute triangle. For this problem, the contestant will have to demonstrate the proof using the obtuse triangle. The basis for the proof is to use the medial triangle. Almost obviously, a perpendicular bisector of $\triangle ABC$ is an altitude of the medial $\triangle PQS$. Thus the orthocenter of the medial triangle is the circumcenter of the original triangle. Also the centroid of the original triangle is the centroid of the medial triangle. In the diagram, the proof is based on showing $\triangle HCG$ to be similar to $\triangle ORG$. This would mean that $\angle CGH \cong \angle OGR$ and C, G, and R are collinear; H, G, and P would be collinear. To prove the triangles similar, $\overline{OS} \parallel \overline{HC}$, so by alternate interior angles $\angle HCG \cong \angle ORG$. And because the medial triangle is similar to the original triangle, $CH = 2(SG)$ and $HC = 2(OS)$, and thus the two triangles are similar by SAS.

3. This is an exercise in the text, but it is a very difficult one, and it will be difficult for a student to discuss it in the short time given in the contest. In the diagram above, focus on the $\triangle OQB$. The contestant may focus on one of the other three comparable triangles. First, $OA = OB = R$. The three relationships that lead to the solution are $OQ^2 = R^2 - \frac{1}{2}AC)^2$, $BQ^2 = \frac{1}{2}(AB^2 + AC^2) + (\frac{1}{2}AC)^2$, and $BQ(OG^2 + QG \cdot GB) = OQ^2(BG) + R^2(QG)$. The other related information necessary is that $BG = 2QG$ and $OH = 3OG$. The rest of the proof is algebraic.

$$\left(OG^2 + \frac{1}{3}BQ \cdot \frac{2}{3}BQ\right) = R^2 - \left(\frac{1}{2}AC\right)^2 \frac{2}{3} + R^2 \frac{1}{3}$$

$$OH^2 = (3OG)^2 = 9\left(R^2 - \frac{1}{6}AC^2 - \frac{2}{9}BQ^2\right) = 9R^2 - \frac{3}{2}AC^2 - 2BQ^2 =$$

$$9R^2 - AB^2 - AC^2 - BC^2$$

Additional questions

1. Show that the three medians of a triangle are obviously concurrent, using Ceva's theorem.

   Answer: Every segment is half a side. When substituted in the equation, the lengths of the sides all divide.

2. What is the "nine-point circle" of a triangle?

   Answer: The feet of the three altitudes, the midpoints of the sides, and the midpoints of the segments from the vertices to the ortho-center all lie on the same circle, called the *nine-point circle*.

PRIME NUMBERS

Reference: Stein, S. *Mathematics: The Man-made Universe*, 3d ed., pp. 13–34. San Francisco: W. H. Freeman and Co., 1976.

Problems

1. Define a prime number, give examples showing a prime and a nonprime number, and use the definition to justify your examples.

2. Let A and B be natural numbers, with $A > B$, and suppose that the natural number D divides both A and B. Prove that D divides $A - B$.

3. Show that there is no largest prime by developing the following argument: Suppose there were a largest prime and call it P. Show that $P! + 1$ is either prime or is divisible by a prime larger than P.

Judges' information

1. The natural numbers with only two distinct divisors are called *prime numbers*. An example of a prime number is 3, because its divisors are 1 and 3. A nonprime number is 6 because its divisors are 1, 2, 3, and 6.

2. Because D divides both A and B, there exist natural numbers Q_1 and Q_2 such that $A = Q_1 D$ and $B = Q_2 D$. But then, $A - B = Q_1 D - Q_2 D = (Q_1 - Q_2)D$. Because $A > B$, $Q_1 - Q_2$ is a natural number and so D divides $A - B$.

3. The idea of the proof is that all the divisors of $P! + 1$ (except 1) must be larger than P because each number less than or equal to P leaves a remainder of 1 when divided into $P! + 1$. Thus either $P! + 1$ is prime or it is a product of primes, each larger than P. Equivalently, one may let D be a prime divisor of $P! + 1$. If $D \le P$ then D divides $P!$. It follows from note 2 above that D then divides the difference $(P! + 1) - (P!) = 1$, which is not possible. Thus $D > P$.

Additional questions

1. Factor 40 into a product of primes. Is this product unique?

Answer: $40 = 2^3 \cdot 5$. This product is unique except for the order.

2. In part 2, it was required that $A > B$. Why is this important?

Hint: If $A \le B$, what happens to $A - B$?

Answer: If $A \le B$, then $A - B$ and $Q_1 - Q_2$ are not natural numbers. Note that the reference book overlooks this important point. The lemma in the book is stated incorrectly—or at least inadequately.

PROBABILITY

Reference: Willoughby, S. *Probability and Statistics*, chap. 3. Morristown, N.J.: Silver Burdett, 1968.

Problems

1. Define independent events and conditional probability. Prove that if A and B are independent events with nonzero probabilities, then the probability that A occurs, given the condition that B has occurred, is equal to $P(A)$.

2. What is the probability that in a family with three children the firstborn child is a boy? If it is known that a family with three children has at least one male child among them, what is the probability that the firstborn child is a boy? (Assume for this question and the next one that the probability of any child's being a male is .5.)

3. It was once noted that most of the astronauts were the first males born in their families. This might not be as suprising as it sounds. Explain how you would calculate that a male is the firstborn male in a family. To develop the problem, you may make the following assumptions about the probabilities regarding the number of children in a particular family:

P(1 child)	=	0.25	P(4 children)	=	0.15
P(2 children)	=	0.35	P(5 children)	=	0.07
P(3 children)	=	0.20	P(6 children)	=	0.02

$$P(k \text{ children}) = 0.01, \text{ for all } k > 6$$

[You do not have to do the actual computations.]

Judges' information

1. This part is directly from the text material. $P(A|B) = P(A \cap B)/P(B)$, and A and B are independent events if and only if $P(A \cap B) = P(A)P(B)$. The proof is a theorem in the text. It is very simple, if the contestant is prepared.

2. The first question in this part assumes that the number of children in a family and the sex of the first child are independent events. Therefore, using the preceding theorem, the probability of the first child's being male is .5. In the second question, the contestant should note that there are seven equally likely cases where there is at least one male in the family of three children. In four of the cases the first child is male. This text tends to use a listing process. Therefore the cases are *bbb, bbg, bgb, bgg, gbb, gbg, ggb*. The probability is then 4/7.

3. This question is both more difficult and open ended. Using the partitioning data given here, which is not necessarily correct, the contestant would have to consider all cases. For this problem, to create a listing there needs to be a consideration for the specific male rather than "a" male being the firstborn. First it needs to be assumed that the number of children in a family and the sex of the children are independent. Therefore, if the astronaut is male, the probability that he is from a family with only one child is .25. Similarly, the probability that he is from a family with more children is given in the table. (For notation, M will be used to represent the specific male considered, and m for other males.) The process would then go as follows:

P(oldest male in a family with one child) = 1.0(.25) = .25

P(oldest male in a family with two children) = (3/4)(.35) = .26

    Cases: *Mm, mM, Mf, fM*

    In three out of four cases he is the oldest male.

P(oldest male in a family with three children) = (7/12)(.20) = .12
    Cases: *Mmm, Mmf, Mfm, Mff, mMm, mMf, fMm, fMf, mmM,*
        mfM, fmM, ffM
    In seven cases out of twelve, he is the oldest male.

If this process were continued with the obvious estimation data for families greater than six, the total probability would be about .73. However, you should be able to determine if the contestant understands the process without having him or her carry out the calculations.

Additional questions

1. In what ways are the probabilities of tossing heads and of having a horse win a race different?

   Answer: The probability of tossing heads is random, and obviously a horse's winning or losing is not a random event.

2. Each night before he goes to sleep, Tom flips a coin. *Heads* means he will wake at 7:00 and *tails* means he will sleep ten minutes longer, to 7:10. During one week he wakes at 7:00 on Monday, 7:00 on Tuesday, 7:00 on Wednesday, and 7:10 on Thursday. What time, if any, is he more likely to wake on Friday?

   Answer: Neither is more likely. The events are independent.

PROJECTIVE GEOMETRY

Reference: Smart, J. R. *Modern Geometrics*, 2d ed., pp. 233–46. Monterey, Calif.: Brooks/Cole Publishing Co., 1978.

Problems

1. Define duality as applied to plane projective geometry and as applied to space geometry. Which of the basic postulates of projective geometry justify the use of dual proofs in plane projective geometry?

2. *a)* Explain why the definitions of a complete quadrangle and a complete quadrilateral are plane duals.

   b) Define a harmonic set of points and a harmonic set of lines, then explain why they are duals.

3. State the dual of the following theorem and prove this dual directly:

   If four points form a harmonic set of points, then the four lines joining these four points to a fifth, noncollinear point are a harmonic set of lines.

Judges' information

1. Generally, duality in a plane allows for the interchanging of the words *point* and *line*, or their applications. For example the sides of a triangle are interchanged with the vertices. In space, the interchange is between points and planes, with lines remaining the same. Axiom 4, the statement *If A, B, C, and D are four distinct points such that $\overleftrightarrow{AB}$ meets $\overleftrightarrow{CD}$, then $\overleftrightarrow{AC}$ meets $\overleftrightarrow{BD}$* implies that there are no quadrilaterals with a pair of parallel sides. This eventually implies that in ordinary Euclidean geometry, when applied to the projective postulates, there are no parallels.

2. In the duals, the definition or theorem will be stated with the dual wording in parentheses.

   a) A complete quadrangle is a set of four points (lines) in a plane, no three are collinear (concurrent), and the lines (vertices) joining (formed by) these vertices (lines) are in pairs.

   This merely defines the quadrangle by its vertices or sides.

   b) A harmonic set of points (lines) is a set of four collinear (concurrent) points (lines) consisting of the four points (lines) of intersection (containing two of the sides of a complete quadrilateral) of the sides of a complete quadrangle with a line passing through two diagonal points (two lines being the diagonal lines of the complete quadrilateral).

3. The dual would be

   If four lines form a harmonic set of lines, then any line not through the point of concurrency will intersect the four lines forming a harmonic set of points.

   The proof might be as follows:

   Let *H(eg, hi)* be a harmonic set of lines and *m* be any nonconcurrent line.

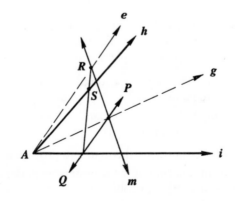

This harmonic set implies the existence of a complete quadrilateral. If *e*, *g* are diagonal lines and *i*, *h* are sides of a quadrangle, then there is a quadrangle *RQAS*. (There could be others by choice of *P*, *Q*). Therefore, the points satisfy the definition of a harmonic set of points.

Additional questions

1. Which of the following are invariant in projective geometry: distance between points, ratio of distance between points, ratios of areas?

   Answer: Distances are not invariant, but ratio of distances and ratio of areas are invariant.

2. Which of the five axioms for projective geometry are also axioms for Euclidean geometry?

   Answer: Two points lie on exactly one line. There are at least four points of which no three are collinear. The three diagonal points of a complete quadrangle are never collinear.

Power Team Using Two, Three, or More Students

A power team is very different from team competitions discussed earlier. Like the oral competition, the mathematics for this competition is more sophisticated than the mathematics in the other competitions. The purpose of this competition is to give the most talented students an opportunity to solve complex problems. As in the oral competition, topics are identified, but specific references need not be given. For example, a topic might be "Conic Sections." The questions posed are related to conic sections, but the problem should·be extremely difficult, as you will see in the examples.

Teams consist of a fixed number of students from each school. The league central committee can set the size of the teams. The teams may be given one or two problems from topics previously identified. Each problem may consist of several parts, with the parts weighted by difficulty. The contestants may have up to forty-five minutes to solve the problems. They may work together, disucssing the problems, if they choose.

Their answer will then be judged on the basis of the correctness and completeness of the solution. Some credit can be given to neatness and organization. Each problem will be worth 25 points, and a team can thus earn a maximum of 50 points.

The difficulty with this competition, as with the oral competition, is the subjectivity of the evaluation. To make scoring more reliable, each solution should be judged independently by two judges, the average of the two scores becoming the contestants' score for the competition. As in judging the oral competition, judges should have a preliminary meeting to discuss the criteria for grading a solution.

Only two examples are given in this category, for the question writer should also be the judge. Because there is no specific reference, solutions to the problems may vary considerably. The two examples differ slightly. The first is worth 25 points, but the first part is worth more than the second. In the second example the second and third parts are equally weighted. In the first example much more information is given to the judges. In the second only the answers are given because there will probably be considerable variation in the methods of proof.

NUMBER THEORY

1. (8 points) State and prove divisibility theorems for 9 and 11. (4 points each)

2. (7 points) Consider the following divisibility theorem for 7:

An integer is divisible by 7 if and only if the number determined by subtracting twice the units from the number with the units digits omitted is divisible by 7.

Demonstrate the theorem by reducing 15 169 to 7. Prove the theorem.

3. (10 points) Determine a divisibility test for 13. Demonstrate the theorem for 3 618 784 and prove it.

Judges' information

1. The "better" proofs are those for which the contestants use generalized expansions of the numbers. That is,

$$N = a_n 10^n + a_{n-1} 10^{n-1} + \cdots + a_1 10^1 + a_0.$$

These two proofs are not exceptionally difficult. It is primarily a matter of rewriting the expanded form in a useful way.

[For 9] $N = a_n(10^n - 1 + 1) + a_{n-1}(10^{n-1} - 1 + 1)$
$$+ \cdots + a_1(10 - 1 + 1) + a_0$$
$$= a_n(10^n - 1) + a_{n-1}(10^{n-1} - 1) + \cdots + a_1(10 - 1)$$
$$+ (a_n + a_{n-1} + \cdots + a_1 + a_0).$$

$10^k - 1 = \overset{k-9}{\overline{999 \ldots 9}}$. Therefore, N is divisible by 9 if and only if $(a_n + a_{n-1} + \cdots + a_0)$ is divisible by 9.

[For 11] $N = a_n(10^n + (-1)^{n+1} + (-1)^n) + \cdots$
$$+ a_2(10^2 - 1 + 1) + a_1(10 + 1 - 1) + a_0$$
$$= a_n(10^n + (-1)^{n+1}) + a_2(10^2 - 1) + a_1(10 + 1)$$
$$+ (a_n(-1)^n + a_n - a_1 + a_2)$$

Because $10^n + (-1)^{n+1}$ is divisible by 11, which can be shown using several approaches, N is divisible by 11 if and only if $(a_n(-1)^n + \cdots + a_2 - a_1 + a_0)$ is divisible by 11.

2. In some ways, this problem is easier than the first two. To show the example, the sequence of new numbers formed by the procedure is

$$15\ 169 \to 1\ 516 - 18 = 1\ 498$$
$$1\ 498 \to 149 - 16 = 133$$
$$133 \to 13 - 6 = 7$$

To prove this, the contestant will have to use an expanded form for the original and new numbers.

$$N = a_n 10^n + a_{n-1} 10^{n-1} + \cdots + a_1 10^1 + a_0$$
$$M = N - (20a_0 + a_0) = N - 21a_0 \text{ or}$$
$$N = M + 21a_0$$

Hence, the proof is simple. M is divisible by 7 if N is divisible by 7, and conversely.

3. There is flexibility in this part because some divisibility theorems can be more sophisticated than others. A simple one is based on the fact that 1 001 is divisible by 13. Hence, discarding the unit digit and subtracting the unit digit times 100 will generate a new number that is congruent to the original number modulo 13. For example: Is 3 618 784 divisible by 13?

$$
\begin{aligned}
3\ 618\ 784 &\rightarrow 361\ 878 - 400 = 361\ 478 \\
361\ 478 &\rightarrow\ \ 36\ 147 - 800 =\ \ 35\ 347 \\
35\ 347 &\rightarrow\ \ \ \ 3\ 534 - 700 =\ \ \ \ 2\ 834 \\
2\ 834 &\rightarrow\ \ \ \ \ \ \ 283 - 400 =\ \ \ \ -117\ .
\end{aligned}
$$

The number -117 is divisible by 13. The proof is much like the proof for divisibility by 7: $M = N - 1001a_0$.

SOLID GEOMETRY

1. (5 points) Find the volume of a regular tetrahedron having six edges of length x. Justify your answer.

2. (10 points) Find the volume of a regular tetrahedron having five edges of length x and one edge of length y. Justify your answer.

3. (10 points) Find the radius of a sphere that circumscribes the tetrahedron in part 2. Justify your answer.

Judges' information

For each of the three parts only the eventual answer will be given. The solution will have to be judged according to the extent of justification.

1. $\dfrac{x^2}{12}\sqrt{2}$

2. $\dfrac{xy}{12}\sqrt{3x^2 - y^2}$

3. This part is easier algebraically but may be more difficult conceptually.

$$
R = \frac{2x}{4x^2 - y^2}\sqrt{3x^4 - 12x^2y^2 + y^4}
$$

ANSWERS

Written Competition
Open Categories, Questions Unweighted

ALGEBRA

1) -12

2) 211

3) 40

4) $\dfrac{1}{2}$

5) 4

6) 36

7) 12

8) 2

9) 18

10) 10.654

11) 1 hr.

12) 4

13) all reals

14) $\dfrac{5m}{4m + 9n}$

15) $\dfrac{-3}{2}$, 1

16) 7.5; 17) $180\sqrt{2}$

18) 7

19) $7{:}05\dfrac{5}{11}$

20) $4m + 2$

GEOMETRY

1) True

2) $\dfrac{3}{\pi}$

3) $121\sqrt{10}$

4) $54°$

5) $4\sqrt{3}$

6) $\sqrt{3}$

7) 13

8) $20\dfrac{5}{8}$

9) $\dfrac{9}{5}$

10) $30°$

11) $32\pi - 64$

12) 14

13) $37 + 3\sqrt{41}$

14) $2 - \sqrt{2}$

15) $\dfrac{45}{2}$ ft.

16) $\dfrac{32\sqrt{2}}{3}$

17) $\dfrac{22}{3}$

18) $\dfrac{32}{5}$

19) 36

20) $14\sqrt{3} - 12$

ADVANCED ALGEBRA

1) 22

2) $\dfrac{8}{5}$

3) 18 ft.

4) 1 080 000

5) 3

6) (0, 6)

7) -8

8) $\dfrac{2x - 3}{2x + 3}$

9) $396k - 99$

10) 75

11) $-2, 3$

12) 19

13) 7

14) 2

15) 8

16) 32

17) 5760

18) 6

19) $\dfrac{992}{3125}$

20) $\dfrac{16}{27}$

PRECALCULUS

1) $-2, 6$

2) $\dfrac{-7}{4}$

3) 1

4) 23

5) $\dfrac{12}{13}$

6) $\dfrac{1}{32}$

7) $\emptyset$

8) $-4i$

9) $\dfrac{50}{111}$

10) $xy = -2$

11) 27

12) $-6 + 3\sqrt{3}$

13) 9

14) 720

15) 0.64

16) $\dfrac{(6 + 4\sqrt{5})}{15}$

17) $2y^2 + 2^2 = 16$

18) $x^2 + 4y^2 = 17$

19) $\sqrt{3} - 1$; 20) 5

Written Competition
Topics Identified, Questions Weighted

FRESHMAN LEVEL

Factoring

1) $3(x + 2)(x - 2)(x^2 + 4)$
2) $(x + 3y)(6xy - 5)$
3) $(x - 4)(50x - 20)$
4) $(m - 4)(m - 3)(m + 1)$
5) -11

Factors and primes

1) 180 000
2) 6 and 60
3) 99 900
4) $(2^4)(3^6)(5^2)(7^2)(3^2)$
5) 31

Logic puzzles

1) D
2) 6
3) Holstein
4) 9
5) Carol

Modular arithmetic

1) 5
2) 0, 4
3) 4
4) Monday
5) 1

Number bases

1) 624
2) 6
3) 110
4) $\dfrac{67}{96}$
5) 3, 4, 6, 5

Rational arithmetic

1) $\dfrac{17}{37}$
2) 318.741
3) 6.204
4) 1.303
5) 10.4%

Word problems

1) $\dfrac{5}{3}$ kg
2) 38
3) $\dfrac{40}{7}$ km/h
4) 841
5) 100 m

Sets and Venn diagrams

1) False
2) 32
3) 335
4) 15
5) 50

SOPHOMORE LEVEL

Arithmetic and geometric progressions

1) 34 650
2) 4 and 16
3) 4.8
4) 14 m
5) $\dfrac{1023}{1024}$

Circles

1) 30°
2) 12 cm
3) $200 - 50\pi$ cm²
4) 20
5) $\dfrac{171}{8}$

Coordinate geometry

1) 1315
2) $2\sqrt{17}$
3) 9
4) (9, 1)
5) $(3, \dfrac{15}{8})$

Geometry of the right triangle

1) 13
2) 30°
3) 2
4) $\dfrac{51}{5}$
5) $\dfrac{100}{3}$

Polygons and polyhedra

1) Triangles
2) 36°
3) $16\sqrt{2}$
4) 20
5) $\dfrac{32}{3}$ cm³

Perimeters and areas

1) True
2) 24
3) $7\sqrt{3}$
4) $2\sqrt{3}$
5) 48

Systems of equations

1) No
2) $\left(3, \dfrac{38}{7}\right)$
3) (1, 0, −1, 2)
4) (29, 2)
5) $-\dfrac{36}{23}$

Similar triangles

1) 63
2) 30
3) 5
4) $\dfrac{4\sqrt{5}}{5}$
5) $\dfrac{36}{5}$

Word problems

1) 15
2) 33
3) (6, 3) and (−6, −3)
4) 32
5) Celia, 4 hrs.

JUNIOR LEVEL

Complex numbers

1) $\dfrac{5i\sqrt{6}}{6}$
2) $\dfrac{-iv\sqrt{5}}{2}$
3) $3 - 2i, -3 + 2i$
4) $-1, 3$
5) $2i\sqrt{7}, -2i\sqrt{7}$

Conic sections

1) (1, 1)
2) C
3) $\dfrac{3}{2}$
4) $(100 - 50\sqrt{3}$ cm, 50 cm) or (134 mm, 50 cm)
5) $\dfrac{x^2}{20} + 5$

Inequalities

1) True

2) $x < -\dfrac{1}{2}$ or $x > \dfrac{3}{2}$

3) $\dfrac{2}{5}$

4) $-1 < x < \dfrac{3}{2}$

5) $-3 < x < -1$,

   $1 < x < \dfrac{53}{37}$

   or $\dfrac{53}{37} < x < 3$

Lines

1) $12\dfrac{3}{8}$

2) 70

3) $-12\dfrac{1}{2}$

4) $\left(\dfrac{5}{4}, \dfrac{73}{20}\right)$

5) $y = (2 + \sqrt{2})x + (4 + 4\sqrt{2})$

Similarity

1) No

2) 8

3) 10

4) $\dfrac{90}{11}$

5) $\dfrac{15}{2}$

Surface areas and volumes

1) $16\sqrt{2}$

2) $\dfrac{96}{\pi^2}$ cm³

3) 18π cm³

4) 78π cm²

5) 1152π

Probability

1) A

2) $\dfrac{132}{425}$

3) $\dfrac{27}{32}$

4) $\dfrac{2}{3}$

5) $\dfrac{9}{160}$

Word problems

1) 61

2) 12

3) 10:20 a.m.

4) $9 - 3\sqrt{5}$

5) 66 hrs.

SENIOR LEVEL

Complex numbers

1) False

2) -1

3) $2\sqrt{2} + 2i\sqrt{2}$, 4(cos 165° + i sin 165°) 4(cos 285° + i sin 285°)

4) $6470i$

5) $\sqrt{3}$(cos 30 + i sin 30) $\sqrt{3}$(cos 150 + i sin 150); $\sqrt{3}$(cos 210° + i sin 210°), $\sqrt{3}$(cos 330° + i sin 330°)

Coordinate geometry

1) False

2) $-\dfrac{1}{2}$ 3

3) $\sqrt{10}$

4) $\dfrac{25}{2}$

5) $4x - 3y = -40$

   or $4x - 3y = 40$

Diophantine equations

1) 7

2) 5

3) 61

4) 11 ft.

5) 84

Logs and exponents

1) $\dfrac{1}{5}$

2) 1.158

3) $x > 4$

4) 64

5) 7

Probability

1) $\dfrac{1}{3}$

2) $\dfrac{1}{2}$

3) $\dfrac{7}{50}$

4) $\dfrac{3}{5}$

5) $\dfrac{27}{100}$

Theory of equations

1) 6

2) 11

3) 36

4) -1

5) $\dfrac{25}{16}$

Trigonometry

1) $\dfrac{\pi}{6}$

2) $\tan \theta$

3) (10, 10, 4)

4) $-4.5\ m$

5) 0.8

Word problems

1) $\dfrac{4}{3}$

2) 15 min

3) 9 hrs.

4) 4 hrs.

5) $\sqrt{74}$

Eight-Person Team Competition

FOUR GRADE LEVELS

1) 12

2) 9

3) 31

4) $3\sqrt{6}$:4

5) 9

6) $\dfrac{1}{3}$

7) $\dfrac{1}{2}$

8) 49

9) 11 520

10) $\dfrac{2\sqrt{3}}{3}$

11) $\sqrt{2} \le y \le 2$

12) 156

13) 58 in.

14) 3

15) Feb. 19, 1873

16) 5 hrs.: $10\dfrac{10}{11}$ min.,

or $10\dfrac{10}{11}$ min. after 5:00

17) -2

18) 2

19) 9

20) $\dfrac{1}{3}$

FRESHMAN-SOPHOMORE LEVEL

1) 680

2) 512

3) $\dfrac{4}{5}$

4) 21

5) $\dfrac{60}{13}$

6) $\sqrt{3}$

7) $145

8) $360

9) $\dfrac{23}{5}$

10) (5, 11)(13, 8)
    (21, 5)(29, 2)

11) $3

12) $61.25

13) C

14) $85

15) 90 sec.

16) 180

17) Cora

18) 9

19) $-\dfrac{3}{2}$

20) 600

JUNIOR-SENIOR LEVEL

1) 7

2) $\dfrac{60}{7}$

3) $\dfrac{48}{55}$

4) $-4, 3$

5) $(10, 3, 1)$

6) -2

7) $\pm(1 + 2\sqrt{5})$

8) $3(\sqrt[3]{18} - \sqrt[3]{9})$ m

9) $\dfrac{4 + \sqrt{37}}{7}$

10) $(6, 6)(-6, -6)$
$(\sqrt{6}, -\sqrt{6})$
$(-\sqrt{6}, \sqrt{6})$

11) 13

12) 0

13) $(5, 20)(6, 12)(8, 8)$

14) $\dfrac{16}{25}$

15) 303

16) 3

17) 3.024

18) $\dfrac{73}{648}$

19) 9

20) $\dfrac{1}{(1 - x)^2}$

Relay Team Competition

FOUR GRADE LEVELS

Round 1

1) 35

2) $300\sqrt{3}$

3) 1

4) $\dfrac{3\pi}{8}$

Round 2

1) 15

2) 75

3) 42

4) -84

Round 3

1) -2

2) $\pi - 2$

3) 14

4) $\dfrac{3}{14}$

FRESHMAN-SOPHOMORE LEVEL

Round 1

1) 104

2) 12

3) $18\sqrt{3}$

4) 3

Round 2

1) $\dfrac{4}{3}$

2) 75

3) 15

4) 6

Round 3

1) 13

2) 7

3) 18

4) 3

Round 4

1) 12

2) 14

3) $7\sqrt{3}$

4) 117

Round 5

1) Any answer

2) 6

3) 6

4) 15

Round 6

1) 51

2) 100

3) 20

4) $10\sqrt{3} + 10$

JUNIOR-SENIOR LEVEL

Round 1

1) 6
2) 9
3) $-\dfrac{9}{2}$
4) $8i$

Round 2

1) 60
2) $20\sqrt{3}$
3) 16
4) 4

Round 3

1) 6
2) 2
3) $2 + \sqrt{2}$
4) 4

Round 4

1) 2
2) -2
3) -4
4) 6.64

Round 5

1) 13
2) 3
3) $\dfrac{11}{12}$
4) $\dfrac{24}{25}$

Round 6

1) 100π
2) 21
3) -21
4) $\dfrac{-24}{25}$

Two-Person Team Competition

FRESHMAN-SOPHOMORE LEVEL

1) 3^{18}
2) 6
3) 37
4) 198
5) 16
6) 5
7) 50
8) $2\sqrt{3}$
9) 22.5 min.
10) 1
11) 13:9
12) 2
13) y
14) 18
15) 3
16) 410
17) 20%
18) 23
19) 5
20) $\dfrac{25}{8}$

JUNIOR-SENIOR LEVEL

1) 6
2) 24:1
3) -9
4) $-\dfrac{12}{13}$
5) 600 m
6) $-\dfrac{8}{9}$
7) 1
8) 6
9) $\dfrac{\pi}{6}$
10) $\dfrac{4}{3}$
11) 108
12) 24
13) 4:1
14) $39\sqrt{3}$
15) $\dfrac{2\pi}{5}$
16) 22
17) $\dfrac{2}{3}$
18) $\dfrac{1}{6}$
19) 18
20) 120

21) 25   25) − 13
22) 2   26) 8   29) $\dfrac{781}{333}$
23) 68   27) 5
24) 47   28) 3   30) − 2

Calculator Competition

1) 3.009   11) 0.5958   21) 10.83 cm
2) − 0.3543   12) 1.047   22) 1.133
3) 4.472   13) − 16.04   23) 0.000 743 6
4) 2.072   14) 10.33   24) − 2.219
5) 3.019   15) 553.5   25) .061 12
6) 1.414   16) 3.562   26) 1.667
7) 1.841   17) 3.141 666 4   27) − 20 590
8) 0.6696   18) 29.2481   28) 0.003 434
9) 6.329   19) − 0.330 82   29) 176.2
10) 0.9341   20) 0.9046   30) 9 662 400 newtons

Estimation Competition

FRESHMAN-SOPHOMORE LEVEL

1) 8.78×10^3 8) 2.87×10^3 15) 8.62 sec.
2) 1.23×10 9) 7.96×10^{-1} 16) 3.21×10^2
3) 7.54   10) 1.91   17) 1.04×10^3 m^3
4) $1.96 \times 10_4$ 11) 1.95   18) 1.14×10^5 yrs.
5) 6.66   12) $9.18   19) 9.85×10^2m^3
6) 1.10×10 13) 2.18×10^9 20) 6.88×10^{-1}
7) 2.13×10 14) 2.18×10^2

JUNIOR-SENIOR LEVEL

1) 2.25×10^{-1} 8) 8.25×10 15) 1.27×10
2) 1.671   9) 7.86×10^{-1} 16) 1.83
3) 1.06×10^7 10) 3.17×10 mph   17) 8.66×10^{-1}
4) 1.20×10^{-2} 11) 1.50×10 18) $−9.41 \times 10^{-1}$
5) 2.74   12) 2.03×10 19) 1.54
6) 1.14   13) 9.76   20) 5.61
7) 8.34×10^{-5} 14) 4.60×10

Multiple-Choice Competition

1) D	14) D	27) B
2) B	15) C	28) B
3) C	16) C	29) E
4) C	17) D	30) E
5) D	18) B	31) C
6) B	19) E	32) D
7) D	20) C	33) A
8) B	21) A	34) D
9) E	22) B	35) C
10) C	23) B	36) D
11) B	24) A	37) E
12) D	25) A	38) B
13) A	26) D	39) B
		40) B